D1223284

Basic Concepts of Synthetic Differential Geometry

Kluwer Texts in the Mathematical Sciences

VOLUME 13

A Graduate-Level Book Series

Basic Concepts of Synthetic Differential Geometry

by

René Lavendhomme
Université Catholique de Louvain

KLUWER ACADEMIC PUBLISHERS
DORDRECHT / BOSTON / LONDON

A C.I.P. Catalogue record for this book is available from the Library of Congress.

ISBN 0-7923-3941-X

Published by Kluwer Academic Publishers,
P.O. Box 17, 3300 AA Dordrecht, The Netherlands.

Kluwer Academic Publishers incorporates
the publishing programmes of
D. Reidel, Martinus Nijhoff, Dr W. Junk and MTP Press.

Sold and distributed in the U.S.A. and Canada
by Kluwer Academic Publishers,
101 Philip Drive, Norwell, MA 02061, U.S.A.

In all other countries, sold and distributed
by Kluwer Academic Publishers Group,
P.O. Box 322, 3300 AH Dordrecht, The Netherlands.

Printed on acid-free paper

All Rights Reserved
© 1996 Kluwer Academic Publishers
No part of the material protected by this copyright notice may be reproduced or
utilized in any form or by any means, electronic or mechanical,
including photocopying, recording or by any information storage and
retrieval system, without written permission from the copyright owner.

Printed in the Netherlands

Table of contents

Table of contents   v

Introduction   xi

1 Differential calculus and integrals   1
   1.1 Introduction : Kock-Lawvere Axiom   1
        1.1.1 Introduction   1
        1.1.2 Notes on logic   2
        1.1.3 A few remarks on infinitesimals   4
        1.1.4 Euclidean R-modules   5
   1.2 The rudimentary differential calculus   6
        1.2.1 The notion of derivative   6
        1.2.2 A Taylor formula   8
        1.2.3 Functions of several variables   11
        1.2.4 Application : a study of homogeneity   15
   1.3 The rudimentary integral calculus   17
        1.3.1 Preorder   17
        1.3.2 Integral on the interval [0, 1]   19
        1.3.3 Integral on an interval [a, b]   21
        1.3.4 A few applications of the axiom of integration . .   24
   1.4 Commented bibliography   30

2 Weil algebras and infinitesimal linearity   33
   2.1 Weil algebras   33
        2.1.1 Introduction   33
        2.1.2 Weil algebras   34
        2.1.3 The general Kock axiom   41

2.2 The general microlinearity of R 43
    2.2.1 The wonderful myopia of R 43
    2.2.2 Quasi colimits of small objects 50
    2.2.3 Perception by R of quasi-colimits of small objects 55
2.3 General microlinearity 57
    2.3.1 The notion of general microlinearity 57
    2.3.2 A few elementary consequences of general micro-
    linearity 57
2.4 Commented bibliography 59

3 Tangency   **61**
3.1 Tangent bundle 61
    3.1.1 The tangent module at a point 61
    3.1.2 Vector bundle and tangent bundle 66
3.2 Vector fields 68
    3.2.1 The R-modules of vector fields 68
    3.2.2 Lie algebra of vector fields 71
3.3 Derivations 76
    3.3.1 Directional derivatives 76
    3.3.2 Reflexive objects 80
3.4 Micro-squares 87
3.5 Commented bibliography 100

4 Differential forms   **101**
4.1 Differential forms with values in an R-module 101
    4.1.1 Singular definition 101
    4.1.2 Classical differential forms 104
4.2 The exterior differential 108
    4.2.1 Infinitesimal n-chains 108
    4.2.2 Integral of a n-form on an infinitesimal n-chain . 110
    4.2.3 The exterior differential 112
4.3 Integral of differential forms 117
    4.3.1 Integral of a n-form on an n-interval 117
    4.3.2 Stokes formula 120
4.4 The exterior algebra of differential forms 123
    4.4.1 Multilinear forms 123
    4.4.2 The exterior product 124

	4.4.3	Interior products. Lie derivatives	126
4.5		De Rham's Theorem	131
	4.5.1	Integration as a cochain morphism	131
	4.5.2	The differentiation morphism D.	132
	4.5.3	Complements	136
4.6		Commented bibliography	139

5 Connections   **141**

5.1		Connection, covariant derivative and spray	141
	5.1.1	Introduction and definition	141
	5.1.2	Covariant derivative	147
	5.1.3	Sprays	151
5.2		Vertical and horizontal microsquares	158
	5.2.1	Vertical-horizontal decomposition	158
	5.2.2	Connecting mappings or connection forms	161
	5.2.3	Parallel transport	164
5.3		Torsion and curvature	166
	5.3.1	Differential forms with value in a tangent fibre bundle	166
	5.3.2	Torsion	169
	5.3.3	Curvature	173
5.4		Commented bibliography	180

6 Global actions   **181**

6.1		Lie objects	181
	6.1.1	Definition of Lie objects	181
	6.1.2	Differential forms	183
	6.1.3	Connections	191
	6.1.4	Extensions of connections and actions	193
6.2		Curvature of a connection on E and torsion of a connection on L	196
	6.2.1	Curvature of a connection	196
	6.2.2	Bianchi identity	200
	6.2.3	Torsion of a connection on L	201
6.3		Weil's characteristic homomorphism	204
	6.3.1	Algebraic preliminaries	204
	6.3.2	A derivation formula	205

6.3.3 The Weil homomorphism 209
6.4 Commented bibliography 210

7 On the algebra of the geometry of mechanics   **211**
7.1 Structured Lie objects 211
    7.1.1 The Riemannian case 211
    7.1.2 Pre-symplectic and symplectic structures 216
    7.1.3 Complex situations 226
    7.1.4 Hermitian and Kaelherian objects 230
7.2 Lie algebras of Lie groups 233
    7.2.1 Lie groups and Lie algebras 239
    7.2.2 Elementar linear examples 241
    7.2.3 Global examples of Lie groups and Lie algebras . 243
    7.2.4 Actions of Lie groups 248
7.3 The cotangent bundle 254
    7.3.1 Vector fields on a vector bundle 254
    7.3.2 Symplectic structure on cotangent bundle 259
    7.3.3 The case of R^n 265
7.4 Commented bibliography 267

8 Note on toposes and models of S.D.G.   **269**
8.1 Multisorted language of higher order 269
    8.1.1 The language 269
    8.1.2 Terms and formulas 271
    8.1.3 Propositional calculus 272
    8.1.4 Predicate calculus 276
    8.1.5 Higher order theory 278
8.2 The concept of topos 279
    8.2.1 (Insufficient) introduction to categories 279
    8.2.2 Definition and examples of toposes 285
    8.2.3 Toposes and set theoretical notions 294
    8.2.4 Language and theory of a topos 297
8.3 Some models of S.D.G. 302
    8.3.1 The algebraic model 302
    8.3.2 Varying sets on $\mathcal{C}^\infty$-algebras 304
    8.3.3 A good model of S.D.G. 308
8.4 Commented bibliography 310

Bibliography   **313**

Index   **317**

Introduction

When thinking about a vector tangent to a surface at some point, geometers readily identify it with a tiny arc of curve having this vector tangent to it. Likewise, when thinking about a vector field, they often like to treat it as being an "infinitesimal transformation", indicating for each point the direction in which the vector field tends to drag it or, alternatively, as a tiny arc along which the field shifts the point during an infinitely small period of time. Finally, all geometers, even if they do not say so explicitly, think of the Lie brackets as being the commutator of the corresponding infinitesimal transformations. But these "ways of thinking" pose a paedagogical problem. If insufficient weight is given to the intuition, differential geometry takes on a rather heavy, technical nature. If, on the other hand, the teacher overemphasizes this intuition, the student may well underestimate the requirement of rigour.

Synthetic differential geometry (S.D.G.), apart from being intrinsically of mathematical interest, provides a new solution to this paedagogical problem. The infinitesimal elements are manipulated explicitly as zero-square elements, giving an accurate content to geometrical intuition and combatting the first threat. These manipulations, however, are carried out in the framework of intuitionistic logic, and experience has shown that the insecurity resulting from unfamiliarity with this logic induces students to maintain sufficient rigour to avoid the second.

The first object of this book is to introduce, in a elementary way, synthetic differential geometry. Our intention is that this text would be readable by graduate students who have not yet taken an advanced course in classical differential geometry. Certainly the reader is not expected to be acquainted with toposes or intuitionistic logic : the constructivist spirit of the text will make the handling of logic easy. While the definitions and theorems have a simple intuitive content, the proofs are usually explicit constructions, taking the form of simple computations.

This book began as (the translation of) a course [41] given to undergraduate students. But afterwards, our aim became more ambitious. Thus we also treat advanced topics, some of them appearing in a book for the first time. These include the treatment of de Rham's theorem at the level of chain complexes, the synthetic view of global action (going as far as the Weil characteristic homomorphism), the systematic account of structured Lie objects (such as Riemannian, symplectic, or Poisson Lie objects), the view of global Lie algebras as Lie algebras of a Lie group in the synthetic sense (e.g. as Lie algebras of Killing vector fields on a Riemannian object), and lastly the synthetic construction of symplectic structure on the cotangent bundle in general (i.e. possibly infinite dimensional case). This last structure, as is well known, leads in the case of R^n to the classical Hamilton equations.

Thus we strictly limit ourselves to a naïve point of view, developing synthetic differential geometry as a theory in itself (within the framework of intuitionistic logic) but nevertheless progressing to somewhat advanced topics, classic in classical differential geometry but new in the synthetic context.

*
*   *

A brief caution for the more informed reader is here apposite: the objects considered correspond not only to the smooth manifolds, but also to those with singularities or of infinite dimension. Thus the elementary results of S.D.G. in this text are sometimes more general than those classically taught. In comparing the classical and synthetic results, we note here that E. Dubuc has shown that there exists a full and faithful embedding of the category of smooth (C^∞ and paracompact) manifolds into a topos, respecting transversal fibre products and open coverings and transforming the real straight line $\mathbb{R}$ into an R-ring of Kock-Lawvere. This is the model introduced briefly in the last chapter.

*

*   *

In the first chapter, we talk about differential calculus and rudimentary integrals. This subject is obviously well-known to the reader in its classical formulation, at least in the finite dimensional case. This chapter contains more than is really needed for what follows, but we thought that it might provide an easy first contact with the spirit of synthetic differential geometry and, as such, merits some development in spite of its elementary character.

The second chapter, on the other hand, may look new to the non-experienced reader. Gathered there are the preliminary algebraic technicalities which lead to the central concept of microlinear objects, whose role in synthetic differential geometry is similar to that of manifolds in classical geometry.

Actual differential gometry starts in Chapter 3 with the notion of "tangency". We develop therein such concepts as tangent fibre bundles and the Lie-algebra of vector fields, as well as directional derivatives. The chapter ends with the description of micro-square space structures.

Chapter 4 studies the exterior algebra of differential forms ; we introduce the notion of exterior differentials as well as form integrals. The Stokes formula is established as in Kock's book [31]. We also include a proof (attributed to Y. Felix and the author) of de Rham's theorem from the point of view of chain complexes.

Chapter 5 is dedicated to affine connections. We establish the equivalence with the definition in terms of covariant derivative and, in the symmetrical case, with the definition in terms of sprays. We develop the themes of (infinitesimal) parallel transport and of (Patterson's) connecting mapping. We also define differential forms with value in a tangent fibre bundle equipped with a connection and their exterior covariant differential. This allows us to give a description of torsion and curvature forms and to establish essential formulas concerning the torsion and curvature tensors.

In Chapter 6 we start with a notion of "Lie object" consisting essentially of an associative algebra together with a Lie algebra of derivations on it, so abstracting the structure of the associative algebra of functions on a manifold together with the Lie algebra of vector fields on it. We

develop there the theory of differential forms, and of connections as
global operators in the manner used classically by Koszul. This is a
good opportunity to recall the Cartan formulas. The chapter closes
with a description of the Weil characteristic homomorphism.

Chapter 7 is, as indicated by its title, directed toward "the algebra
of the geometry of mechanics". First we present various structured Lie
objects: Riemannian, pre-symplectic and symplectic, Poisson, almost
complex and complex, Hermitian and Kaelherian objects. Then, we
develop a systematic treatment of Lie algebras of Lie groups. In S.D.G.
the global (infinite-dimensional) Lie algebras are also Lie algebras of
a Lie group. Thus, while classically the set of diffeomorphisms of a
manifold is not exactly a Lie group , in S.D.G. it is well known that
the corresponding object actually is a Lie group and the set of vector
fields is the Lie algebra of that Lie group. It is interesting to consider
other global Lie algebras: the Lie algebra of Killing vector fields in the
Riemannian case, the Lie algebra of locally Hamiltonian vector fields
in the symplectic case, the Lie algebra of affine vector fields in the
case of an affine global connection, the Lie algebra of Poisson vector
fields in the case of Poisson object. Lastly, we explore the cotangent
bundle where, in the presence of Riemannian structure and connection,
the Liouville forms are easily handled. We prove that the Liouville 2-
form is symplectic; we then deduce, in the case of $\mathbb{R}^n$, the Hamiltonian
equations.

The last chapter is of a different kind. We try to explain why the
naïve set theory used is not pure nonsense by introducing the reader
to intuitionistic logic, and the resulting set theory, from an elementary
point of view. We also introduce the concept of toposes and the most
important model of synthetic differential geometry. We make no at-
tempt to give complete proofs, but we refer the interested reader to the
books of A. Kock [31] , or that of I. Moerdijk and G.E. Reyes [52].

It goes without saying that we cannot seriously hope to make pro-
gress in synthetic differential geometry without a formation in classical
differential geometry; as a symbolic aim, we think appropriate to en-
courage the beginner to look at classical works such as "Foundations of
differential geometry" of S. Kobayashi and K. Nomizu [24] or "A comprehensive introduction to differential geometry" of M. Spivak [55].

*

*   *

Finally I would like to express my gratitude to my colleagues Francis Borceux, Yves Felix, Patrick Habets and Thierry Lucas, who kindly read over parts of the manuscript; I thank the final-year students in mathematics at the Université de Louvain who have taken this course; and finally I thank Béatrice Huberty who was responsible for typing a first version of the work.

I should also express gratitude to Francis Borceux for proposing the translation of the original French of this book, finding a translator, and obtaining financial support for this first translation. Then, following a suggestion of Marta Bunge, I extended the book in the present form. I extend thanks to Stephen Lack for revising the English in this final version, and to Peter Johnstone for referring him to me.

Chapter 1

Differential calculus and integrals

1.1 Introduction : Kock-Lawvere Axiom

1.1.1 Introduction

Mathematicians or physicists often speak, with some neglect, of first order approximation, as if they were manipulating numbers d so small that they have null square.

But if, ultimately, this could be considered as not being a neglect, if the model of the physical straight line could be a ring including elements with square zero, would that not be a considerable simplification ?

A. Grothendieck (cf. for example [22]) insisted on not excluding nilpotent elements in algebraic geometry. To him they even seemed indispensable in the elaboration of a differential calculus adapted to the needs of algebraic geometry.

A decisive step has been achieved by F.W. Lawvere in a series of lectures given in 1967 (cf. [44]). In them he provided an axiom dealing with the set D of elements of square zero in a ring R modelling the straight line. If this axiom is accepted, every function from R to R becomes "differentiable" and thus infinitely differentiable (smooth). From there, Lawvere lays the foundation of a differential geometry rehabilitating intuitive ways of reasoning which employed an apparently vague notion of infinitesimal.

1

Let us now state Lawvere's axiom in the way it has been formulated by A. Kock (cf. [27]). Let R be a unitary commutative ring. We do not suppose that R is a field. But we shall suppose that the elements of R can be divided by 2, 3, ... In other words R is an algebra over $\mathbb{Q}$. Let

$$D = \{d \in R \mid d^2 = 0\}.$$

The Kock-Lawvere axiom postulates that every function from D into R is in a unique way a first-degree function. A. Kock repeats this in a picturesque way : D is so small that one cannot distinguish the graph of a function from D to R from a segment of a straight line, but D is so big that its slope is uniquely determined.

More explicitly, let us put :

(Kock-Lawvere axiom)(K-L). *For every $f : D \to R$, there exists one and only one $b \in R$, such that for every d in D*

$$f(d) = f(0) + d \cdot b.$$

1.1.2   Notes on logic

The logic that will be used in this text is intuitionistic, not classical logic.

We must of course, from the beginning, indicate why the Kock-Lawvere axiom is classically unacceptable :

Catastrophic-proposition. *If R satisfies the Kock-Lawvere axiom, then R is the null ring .*

Proof:

1. Let us prove first that $D = \{0\}$. Indeed let us consider the function $f : D \to R$ described by $f(0) = 0$ and $f(\delta) = 1$ for every non zero δ . There exists then $b \in R$ such that, for every d, $f(d) = d \cdot b$. If there were a non zero $\delta \in D$, we would have $1 = \delta \cdot b$ and thus, multiplying by δ, $\delta = 0$ (since $\delta^2 = 0$). This contradiction shows that $D = \{0\}$.

2. But then $R = \{0\}$ because if the zero function on $D = \{0\}$ is written in a unique way as $0 = 0.b$, it means that every b of R is null. ∎

Thus the theory seems to have stopped before it started. The pessimists ask ironically : how can you expect a function that vanishes at zero and equals to 1 elsewhere to be differentiable at 0 ?

In folklore we find the following witty slogan : if there is a contradiction between physics and logic, you must change the logic.

Fortunately, the logic that we need, weaker than classical logic and allowing less deductions, already exists, and is called intuitionistic logic. It is essentially a logic without the law of the excluded middle. We cannot affirm "*P or not P*" for every proposition *P*. Thus, we do not have the double negation principle :

$$not \ (not \ P) \rightarrow P.$$

But in the previous proof we did make use of the double negation principle. We also used the principle of excluded middle when describing $f : D \rightarrow R$ since by saying $f(0) = 0$ and $f(\delta) = 1$ for $\delta \neq 0$, we implicitly considered as evident that

$$d = 0 \ or \ d \neq 0$$

which is an application of the excluded middle. Without this principle we cannot ensure that this function f is well defined. In particular, if it does not exist the problem of its differentiability is not a matter of discussion !

It is clear that just the fact of remarking that the proof of the "catastrophic-proposition" makes use of non-intuitionistic laws of logic is not sufficient to ensure the good functioning of the theory.

One should, from a positive point of view, have a model. However, this would require revising the classical notion of set-model (as the logic validated by the classical notion of set-model is classical logic itself). The concept of model should therefore be extended to categories other than the category of sets.

That this is possible has been established : E. Dubuc [14] has even been able to construct a model well-adapted to the needs of differential geometry.

But we shall not start there ; we prefer to start our work "as if" we were in the normal context of sets, but forcing ourselves to use only the rules of intuitionistic logic. The main rules of this logic are summarized in the last chapter, and the courageous reader could check the fact that we have used only these rules. For an account of models we refer the reader to more advanced books such as [4] and [5]. A short description of Dubuc's model is given the last chapter.

1.1.3   A few remarks on infinitesimals

The elements d of D are called infinitesimals. These elements cannot be cancelled since

$$d \cdot 0 = d \cdot d.$$

However we do have the following result :

Proposition 1. *Let $a, b \in R$. If, for any d of D, $d \cdot a = d \cdot b$ then $a = b$.*

Proof: This follows immediately from the uniqueness assertion in the axiom (K-L). ∎

Notice that if we suppose $1 \neq 0$ – which is a natural thing to do – one cannot assume D to be an ideal : the sum of two infinitesimals is not necessarily an infinitesimal.

Proposition 2. *D is not an ideal of R.*

Proof: If D is an ideal of R then $\forall d_1, d_2 \in D, d_1 + d_2 \in D$ and thus $0 = (d_1 + d_2)^2 = 2d_1 \cdot d_2$.

One can divide by 2 in R and thus $\forall d_1, d_2 \in D, d_1 \cdot d_2 = 0$. But then according to Proposition 1, $\forall d_2 \in D, d_2 = 0$, hence $D = \{0\}$. But then the zero function in D is written in a single way as $0 = 0 \cdot b$, $R = \{0\}$, and consequently $1 = 0$. Since we assumed $1 \neq 0$, the hypothesis "D is an ideal of R" implies a false statement. Hence D is not an ideal of R. ∎

Notice that the previous reasoning does not appeal to the double negation principle. It uses only that *not P* equals $P \rightarrow false$.

Let us observe that if $d_1, d_2 \in D$, then $d_1 + d_2 \in D$ if and only if $d_1 \cdot d_2 = 0$. Let

$$
\begin{aligned}
D(2) &= \{(d_1, d_2) \mid d_1^2 = d_2^2 = d_1 \cdot d_2 = 0\} \\
&= \{(d_1, d_2) \mid d_1 \in D, d_2 \in D, d_1 + d_2 \in D\}.
\end{aligned}
$$

As D is not an ideal, $D(2) \neq D \times D$. We remark that $D(2)$ includes the axes

$$\{(d,0) \mid d \in D\} \quad \text{and} \quad \{(0,d) \mid d \in D\}$$

but also the "diagonal"

$$\{(d,d) \mid d \in D\}$$

and indeed

$$\{(\alpha d, \beta d) \mid d \in D\}$$

for all α, β of R.

We define more generally

$$D(n) = \{(d_1, d_2, \ldots, d_n) \mid d_i \cdot d_j = 0 \; for \; 1 \le i, j \le n\}.$$

There are other small objects like these. We shall consider them in Chapter 2.

1.1.4 Euclidean R-modules

One might need vector versions of the axiom (K-L). We adopt the following definition.

Definition 1. *Let E be an R-module. We say that E is Euclidean if every function from D to E can be written in a unique way as a first degree function:*

$$\forall f : D \to E \; \exists ! \, \underline{b} \in E \; \forall d \in D \; f(d) = f(0) + d\underline{b}.$$

A few examples follow.

Proposition 3. R^n *is a Euclidean R-module.*

Proof: If $f : D \to R^n$, it suffices to apply the axiom (K-L) to each component of f. ∎

More generally but just as easily :

Proposition 4. *Any product of Euclidean R-modules is a Euclidean R-module. If X is a set and E is a Euclidean R-module, then the R-module E^X of functions from X into E is also Euclidean.*

Notice that, as suggested before, we are using here the set-notions as being just notions of an intuitionistic theory of sets, or, more approprietely, notions of topos theory. We also have

Proposition 5. *If E is Euclidean and V is an R-module, then the R-module $L(V, E)$ is Euclidean.*

1.2   The rudimentary differential calculus

1.2.1   The notion of derivative

Let $f : R \rightarrow R$ and let a be an element of R. Let us consider the function from D into R whose value at d is $f(a + d)$. According to the axiom of Kock-Lawvere, there exists a unique b in R such that

$$\forall\, d \in D \;\; f(a + d) = f(a) + d \cdot b.$$

This element b of R will be denoted $f'(a)$ and is called the *derivative* of f at a.

Thus the Kock-Lawvere axiom implies that every function from R into R is differentiable and consequently that every function from R to R is infinitely differentiable.

Let us show that the derivative that has just been defined possesses the usual elementary properties of the derivative.

Proposition 1. *Let f and g be two functions from R to R and let $\alpha \in R$. We have*

     1. $(f + g)' = f' + g'$

     2. $(\alpha \cdot f)' = \alpha \cdot f'$

     3. $(f \cdot g)' = f' \cdot g + f \cdot g'$.

Thus, the derivation-operation is actually a "derivation" in the usual algebraic sense.

Proof: On the one hand, we have by definition

$$(f + g)(x + d) = (f + g)(x) + (f + g)'(x) \cdot d$$

and on the other hand,

$$\begin{aligned}
(f + g)(x + d) &= f(x) + f'(x) \cdot d + g(x) + g'(x) \cdot d \\
&= (f + g)(x) + (f'(x) + g'(x)) \cdot d
\end{aligned}$$

and thus identity (1) holds. Identity (2) is similar and identity (3) results from the following trivial computation (where we use $d^2 = 0$) :

$$\begin{aligned}
(f \cdot g)(x + d) &= (f(x) + f'(x) \cdot d) \cdot (g(x) + g'(x) \cdot d) \\
&= (f \cdot g)(x) + (f(x) \cdot g'(x) + f'(x) \cdot g(x)) \cdot d.
\end{aligned}$$

∎

Proposition 2. *Let* $R \xrightarrow{g} R \xrightarrow{f} R$. *We have the chain rule for the derivative of a composite :*

$$(f \circ g)' = (f' \circ g) \cdot g'.$$

Proof: On the one hand

$$f(g(x + d)) = f(g(x)) + (f \circ g)'(x) \cdot d$$

and on the other hand :

$$\begin{aligned}
f(g(x + d)) &= f(g(x) + g'(x) \cdot d) \\
&= f(g(x)) + f'(g(x)) \cdot g'(x) \cdot d
\end{aligned}$$

by definition of f' since $g'(x) \cdot d \in D$. Thus we obtain the announced formula. ∎

Notice that if E is a Euclidean R-module, then the formula $f(x+d) = f(x)+f'(x) \cdot d$ allows us in the same way to differentiate every function from R into E. We easily get the elementary algebraic properties of this notion of derivative . As an example let us prove Leibniz's law for an arbitrary bilinear product.

Proposition 3. *Let E, F and G be Euclidean R-modules and let $m :$ $E \times F \to G$ be a bilinear mapping. Let $f : R \to E$ and $g : R \to F$. Defining $f \underset{m}{\cdot} g : R \to G$ by means of m (i.e. $f \underset{m}{\cdot} g = m \circ (f, g)$), we have*

$$(f \underset{m}{\cdot} g)' = f' \underset{m}{\cdot} g + f \underset{m}{\cdot} g'.$$

Proof: This results from the following computation for each x of R and d of D :

$$(f \underset{m}{\cdot} g)(x + d) = f(x + d) \underset{m}{\cdot} g(x + d)$$
$$= (f(x) + f'(x).d) \underset{m}{\cdot} (g(x) + g'(x).d)$$
$$= f(x) \underset{m}{\cdot} g(x) + (f'(x) \underset{m}{\cdot} g(x) + f(x) \underset{m}{\cdot} g'(x)).d.$$

∎

1.2.2   A Taylor formula

The formula

$$f(x + d) = f(x) + f'(x) \cdot d$$

can be considered as a Taylor formula. As $d_1 + d_2$ is not necessarily in D, we get a formula for higher orders by applying several times the previous formula :

$$f(x + d_1 + d_2) = f(x + d_1) + d_2 f'(x + d_1)$$
$$= f(x) + d_1 f'(x) + d_2(f'(x) + d_1 f''(x))$$
$$= f(x) + f'(x) \cdot (d_1 + d_2) + f''(x) \cdot d_1 \cdot d_2$$

or :

$$f(x + d_1 + d_2 + d_3) = f(x) + f'(x)(d_1 + d_2 + d_3)$$
$$+ f''(x)(d_1 d_2 + d_1 d_3 + d_2 d_3) + f'''(x) \cdot d_1 d_2 d_3.$$

We thus find, as coefficients of the derivatives, the elementary symmetric polynomials. We denote by σ_k^n for $1 \le n \le k$ these elementary symmetric polynomials ; $\sigma_k^n(X_1, \ldots, X_k)$ is the sum of the products of n distinct factors. For instance

$$
\begin{aligned}
\sigma_k^1(X_1, \ldots, X_k) &= X_1 + X_2 + \ldots + X_k \\
\sigma_k^2(X_1, \ldots, X_k) &= \sum_{i \neq j} X_i X_j \\
\sigma_k^k(X_1, \ldots, X_k) &= X_1 X_2 \ldots X_k.
\end{aligned}
$$

One often puts $\sigma_k^0 = 1$ and $\sigma_k^n = 0$ for $n > k$. Notice also the inductive formula :

$$
\sigma_{k+1}^n(X_1, \ldots, X_{k+1}) = \sigma_k^n(X_1, \ldots, X_k) + X_{k+1} \sigma_k^{n-1}(X_1, \ldots, X_k).
$$

We can now state a Taylor formula :

Proposition 4. *For every function $f : R \to R$, for every x in R and $d_1, \ldots, d_k$ in D, we have*

$$
f(x + d_1 + \ldots + d_k) = \sum_{n=0}^{k} f^{(n)}(x) \sigma_k^n(d_1, \ldots, d_k).
$$

Proof: The formula being immediate for $k = 1$, one may proceed by induction on k. But since

$$
f(x + d_1 + \ldots + d_k + d_{k+1}) = f(x + d_1 + \ldots + d_k) + d_{k+1} \cdot f'(x + d_1 + \ldots + d_k)
$$

we apply the inductive assumption to f and f' and so :

$$
\begin{aligned}
f(x + d_1 &+ \ldots + d_k + d_{k+1}) \\
&= \sum_{n=0}^{k} f^{(n)}(x) \sigma_k^n(d_1, \ldots, d_k) \\
&\quad + d_{k+1} \sum_{n=0}^{k} f^{(n+1)}(x) \sigma_k^n(d_1, \ldots, d_k) \\
&= \sum_{n=0}^{k+1} f^{(n)}(x)(\sigma_k^n(d_1, \ldots, d_k) + d_{k+1} \sigma_k^{n-1}(d_1, \ldots, d_k)) \\
&= \sum_{n=0}^{k+1} f^{(n)}(x) \sigma_k^n(d_1, \ldots, d_k, d_{k+1}).
\end{aligned}
$$

Thus the proposition has been proved. ∎

To recover the previous formula in its more usual form, it suffices to use the following lemma :

Lemma. *For all k and all n :*

$$(d_1 + d_2 + \ldots + d_k)^n = n! \sigma_k^n(d_1, \ldots, d_k).$$

Proof: Let us use induction on k. We have, since $d_{k+1}^2 = 0$,

$$(d_1 + \ldots + d_k + d_{k+1})^n = (d_1 + \ldots + d_k)^n + n d_{k+1}(d_1 + \ldots + d_k)^{n-1}$$

which gives, by the inductive assumption :

$$n! \, \sigma_k^n(d_1, \ldots, d_k) + n.d_{k+1}.(n-1)! \sigma_k^{n-1}(d_1, \ldots, d_k)$$
$$= n! \, (\sigma_k^n(d_1, \ldots, d_k) + d_{k+1}\sigma_k^{n-1}(d_1, \ldots, d_k))$$
$$= n! \, \sigma_{k+1}^n(d_1, \ldots, d_k, d_{k+1}).$$

■

Proposition 4 and the lemma give :

Proposition 5. *For every function* $f : R \to R$, *for every* x *of* R *and for every* δ *of* R *having the form* $\delta = d_1 + \ldots + d_k$ *with each* d_i *in* D, *one has*

$$f(x + \delta) = \sum_{n=0}^{k} f^{(n)}(x) \cdot \frac{\delta^n}{n!}.$$

One also has the uniqueness result :

Proposition 6. *Let* $f : R \to R$. *If for each* $d_1 \ldots d_k$ *of* D *we have :*

$$
\begin{aligned}
f(x + d_1 + \ldots + d_k) \;=\;& a_0 + \sum_{i=1}^{k} a_{1,i} d_i \\
& + \sum_{i<j} a_{2,ij} d_i d_j + \ldots \\
& + \sum_{\substack{i_1,\ldots,i_r \\ 1 \le i_1 < \cdots < i_r \le k}} a_{r,i_1 i_2 \ldots i_r} d_{i_1} \ldots d_{i_r} \\
& + \ldots + a_{k,1\,2\ldots k} d_1 d_2 \ldots d_k,
\end{aligned}
$$

then

$$a_{r,i_1 i_2 \ldots i_r} = f^{(r)}(x).$$

Proof: Letting $d_1 = d_2 = \ldots = d_k = 0$, we obtain $a_0 = f(x)$. Let us proceed by induction on r. Let us show that $a_{r,i_1 i_2 \ldots i_r} = f^{(r)}(x)$. In the equalities of Propositions 4 and 6 we put $d_i = 0$ for i different from $i_1, i_2, \ldots, i_r$. From the inductive assumption and the equality $\sigma_k^r(d_1, \ldots, d_k) = d_{i_1} \ldots d_{i_r}$, one gets

$$f^{(r)}(x) d_{i_1} d_{i_2} \ldots d_{i_r} = a_{r,i_1 i_2 \ldots i_r} d_{i_1} d_{i_2} \ldots d_{i_r}.$$

The equalities being true for all $d_{i_1}, \ldots, d_{i_r}$ of D, one gets indeed

$$f^{(r)}(x) = a_{r,i_1 i_2 \ldots i_r}.$$

∎

1.2.3 Functions of several variables

Let V be an R-module, E an Euclidean R-module and $f : V \to E$.

Let us first define the concept of *directional derivative* . Let $\underline{a}, \underline{u}$ be in V. Consider the function from D to E given by $f(\underline{a} + d\underline{u})$. Since E is Euclidean, there exists a unique $\underline{b}$ in E such that

$$\forall d \in D, \ f(\underline{a} + d\underline{u}) = f(\underline{a}) + d \cdot \underline{b}.$$

This unique $\underline{b}$ is denoted $(\partial_{\underline{u}} f)(\underline{a})$ and is called the derivative of f in the direction $\underline{u}$ at $\underline{a}$.

Proposition 7. *For every $\underline{a}$ of V, $\partial_{\underline{u}} f(\underline{a})$ is a linear function of $\underline{u}$.*

Proof: Let us first show homogeneity. In identity (1) :

$$f(\underline{a} + d\underline{u}) = f(\underline{a}) + d\partial_{\underline{u}} f(\underline{a}) \tag{1}$$

nothing changes if we replace d by $d \cdot \lambda$ or $\underline{u}$ by $\lambda \underline{u}$. Replacing d by $d \cdot \lambda$ in (1), we obtain

$$f(\underline{a} + d \cdot \lambda \cdot \underline{u}) = f(\underline{a}) + d \cdot \lambda \cdot \partial_{\underline{u}} f(\underline{a})$$

and replacing $\underline{u}$ by $\lambda \cdot \underline{u}$, we get

$$f(\underline{a} + d \cdot \lambda \cdot \underline{u}) = f(\underline{a}) + d.\partial_{\lambda \underline{u}} f(\underline{a}).$$

Subtracting one equation from the other, one obtains that for all d

$$d \cdot \lambda \cdot \partial_{\underline{u}} f(\underline{a}) = d \cdot \partial_{\lambda \underline{u}} f(\underline{a})$$

and thus $\partial_{\lambda \underline{u}} f(\underline{a}) = \lambda \cdot \partial_{\underline{u}} f(\underline{a})$. This is the desired homogeneity. Let us now prove additivity. On the one hand

$$f(\underline{a} + d(\underline{u}_1 + \underline{u}_2)) = f(\underline{a}) + d\partial_{\underline{u}_1 + \underline{u}_2} f(\underline{a})$$

and on the other hand

$$
\begin{aligned}
f(\underline{a} + d\underline{u}_1 + d\underline{u}_2) &= f(\underline{a} + d\underline{u}_1) + d\partial_{\underline{u}_2} f(\underline{a} + d\underline{u}_1) \\
&= f(\underline{a}) + d\partial_{\underline{u}_1} f(\underline{a}) + d(\partial_{\underline{u}_2} f(\underline{a}) + d\partial_{\underline{u}_1}(\partial_{\underline{u}_2} f)(\underline{a})) \\
&= f(\underline{a}) + d(\partial_{\underline{u}_1} f(\underline{a}) + \partial_{\underline{u}_2} f(\underline{a}))
\end{aligned}
$$

since $d^2 = 0$. So one finds

$$\partial_{\underline{u}_1 + \underline{u}_2} f(\underline{a}) = \partial_{\underline{u}_1} f(\underline{a}) + \partial_{\underline{u}_2} f(\underline{a}).$$

∎

The linear function that has just been described is called the *differential* of f with respect to $\underline{a}$. We write

$$f'(\underline{a}) = df(\underline{a}) \; : \; V \longrightarrow E$$
$$\underline{u} \rightsquigarrow df(\underline{a})(\underline{u}) = \partial_{\underline{u}} f(\underline{a}).$$

We can verify the usual algebraic rules in the same way as we did with functions of one variable : the linearity of $\partial_{\underline{u}} f$ in $\underline{u}$ or the linearity of df in f, the Leibniz rules in the presence of a bilinear multiplication, or the rules of functoriality of the differential :

$$d(g \circ f)(\underline{a}) = dg(f(\underline{a})) \circ df(\underline{a}).$$

In the case where $V = R^n$, we define the partial derivatives by

$$\frac{\partial f}{\partial x_i}(\underline{a}) = \partial_{\underline{e}_i} f(\underline{a})$$

and we have the formula

$$\partial_{\underline{u}} f(\underline{a}) = \sum_{i=1}^{n} \frac{\partial f}{\partial x_i}(\underline{a}) \cdot u_i$$

if $\underline{u} = (u_1, \ldots, u_n)$. We also have

$$df(\underline{a}) = \sum_{i=1}^{n} \frac{\partial f}{\partial x_i}(\underline{a}) \cdot dx_i$$

where dx_i is the differential of the i-th projection $\pi_i : R^n \to R$.

Higher degree derivatives can also now be introduced. Let $f : V \to E$ where V is an R-module and E an Euclidean R-module. The derivative or differential of f

$$f' : V \to L(V, E)$$

is itself differentiable since $L(V, E)$ is still an euclidean R-module. Let $\underline{a} \in V$. One defines $f''(\underline{a}) : V \times V \to E$ by

$$f''(\underline{a})(\underline{u}, \underline{v}) = \partial_{\underline{v}}(\partial_{\underline{u}} f)(\underline{a}).$$

Proposition 8. *The function*

$$f''(\underline{a}) : V \times V \to E$$

is a symmetric bilinear function.

Proof: The fact that $f''(\underline{a})$ is bilinear follows immediately from Proposition 7 : the directional derivative is linear. The symmetry remains to be proved :

$$\partial_{\underline{v}}(\partial_{\underline{u}} f)(\underline{a}) = \partial_{\underline{u}}(\partial_{\underline{v}} f)(\underline{a}).$$

Let us consider d_1 and d_2 in D. We have

$$f(\underline{a} + d_1\underline{u} + d_2\underline{v}) - f(\underline{a} + d_1\underline{u}) - f(\underline{a} + d_2\underline{v}) + f(\underline{a})$$
$$= d_2 \cdot \partial_{\underline{v}} f(\underline{a} + d_1\underline{u}) - d_2\partial_{\underline{v}} f(\underline{a}))$$
$$= d_2 \cdot (\partial_{\underline{v}} f(\underline{a} + d_1\underline{u}) - \partial_{\underline{v}} f(\underline{a}))$$
$$= d_2 \cdot d_1 \cdot \partial_{\underline{u}}(\partial_{\underline{v}} f)(\underline{a}).$$

Furthermore, the same expression can be written as

$$f(\underline{a} + d_2\underline{v} + d_1\underline{u}) - f(\underline{a} + d_2\underline{v}) - f(\underline{a} + d_1\underline{u}) + f(\underline{a})$$
$$= d_1 \cdot \partial_{\underline{u}} f(\underline{a} + d_2\underline{v}) - d_1\partial_{\underline{u}} f(\underline{a}))$$
$$= d_1 \cdot d_2\partial_{\underline{v}}(\partial_{\underline{u}} f)(\underline{a}).$$

Thus we have

$$d_1 d_2 \partial_{\underline{u}}(\partial_{\underline{v}} f)(\underline{a}) = d_1 d_2 \partial_{\underline{v}}(\partial_{\underline{u}} f)(\underline{a})$$

whence the announced symmetry.   ∎

This process can of course be iterated and allows us to define

$$f^{(k)}(\underline{a}) : V^k \to E$$

$$(\underline{u}_1, \ldots, \underline{u}_k) \rightsquigarrow \partial_{\underline{u}_1} \partial_{\underline{u}_2} \ldots \partial_{\underline{u}_k} f(\underline{a})$$

which is a symmetric k-linear function.

One can adapt the argument of 1.2.2 to find the following Taylor formula with infinitesimal increments.

Proposition 9. *Let* $f : V \to E$, $\underline{a} \in V$ *and* $\underline{h}$ *an increment that can be written as follows*

$$\underline{h} = d_1 \underline{u}_1 + d_2 \underline{u}_2 + \ldots + d_k \underline{u}_k.$$

Then,

$$f(\underline{a} + \underline{h}) = \sum_{n=0}^{k} \frac{1}{k!} f^{(k)}(\underline{a})(\underline{h}, \ldots, \underline{h}).$$

Proof: We shall verify the above proposition only for $k = 2$. We have

$$\begin{aligned}
f(\underline{a} + d_1 \underline{u}_1 + d_2 \underline{u}_2) &= f(\underline{a} + d_1 \underline{u}_1) + d_2 f'(\underline{a} + d_1 \underline{u}_1)(\underline{u}_2) \\
&= f(\underline{a}) + d_1 f'(\underline{a})(\underline{u}_1) + d_2 f'(\underline{a})(\underline{u}_2) + d_1 d_2 f''(\underline{a})(\underline{u}_1, \underline{u}_2) \\
&= f(\underline{a}) + f'(\underline{a})(d_1 \underline{u}_1 + d_2 \underline{u}_2) + d_1 d_2 f''(\underline{a})(\underline{u}_1, \underline{u}_2).
\end{aligned}$$

However

$$f''(\underline{a})(d_1 \underline{u}_1 + d_2 \underline{u}_2, d_1 \underline{u}_1 + d_2 \underline{u}_2) = d_1 d_2 f''(\underline{a})(u_1, u_2) + d_2 d_1 f''(\underline{a})(\underline{u}_2, \underline{u}_1)$$

which by the symmetry of $f''(a)$ is equal to

$$2 d_1 d_2 f''(\underline{a})(\underline{u}_1, \underline{u}_2).$$

Thus we have with $\underline{h} = d_1 \underline{u}_1 + d_2 \underline{u}_2$,

$$f(\underline{a} + \underline{h}) = f(\underline{a}) + f'(\underline{a})(\underline{h}) + \frac{1}{2} f''(\underline{a})(\underline{h}, \underline{h})$$

which is the Taylor formula of Proposition 9 in the case where $k = 2$.

1.2.4 Application : a study of homogeneity

A first application of the linearity of $\partial_{\underline{u}}$ is the linearity of homogeneous mappings. We shall start by proving a lemma. Let, again, V be an R-module, E a Euclidean R-module and $f : V \rightarrow E$. f is said to be *homogeneous* if for all λ of $R : f(\lambda\underline{a}) = \lambda f(\underline{a})$.

Lemma. *If $f : V \rightarrow E$ is homogeneous, then $df(\underline{a})$ is independent of $\underline{a}$ (i.e. for each $\underline{a} : df(\underline{a}) = df(\underline{0})$).*

   Proof: Recall that

$$f(\underline{a} + d\underline{u}) = f(\underline{a}) + d \cdot \partial_{\underline{u}} f(\underline{a}). \tag{1}$$

By applying (1) at the point $\lambda\underline{a}$ and the infinitesimal λd, we find

$$f(\lambda\underline{a} + \lambda.d.\underline{u}) - f(\lambda\underline{a}) = \lambda \cdot d \cdot \partial_{\underline{u}} f(\lambda\underline{a}).$$

As f is homogenous, the left hand side equals $\lambda(f(\underline{a} + d\underline{u}) - f(\underline{a}))$ and we have, for each d of D,

$$\lambda \cdot d \cdot \partial_{\underline{u}} f(\lambda\underline{a}) = \lambda \cdot d \cdot \partial_{\underline{u}} f(\underline{a}).$$

Thus

$$\lambda \cdot \partial_{\underline{u}} f(\lambda a) = \lambda \cdot \partial_{\underline{u}} f(\underline{a}).$$

Differentiating the two members with respect to λ , gives

$$\partial_{\underline{u}} f(\lambda\underline{a}) + \lambda \frac{\partial}{\partial\lambda}(\partial_{\underline{u}} f(\lambda\underline{a})) = \partial_{\underline{u}} f(\underline{a}).$$

By letting $\lambda = 0$, we obtain as announced

$$\partial_{\underline{u}} f(\underline{0}) = \partial_{\underline{u}} f(\underline{a}).$$

■

Proposition 10. *If f is homogeneous,then f is linear and $f = df(\underline{0})$.*

Proof: Putting $\underline{a} = \underline{u}$ in equality (1) yields :

$$
\begin{aligned}
d \cdot \partial_{\underline{u}} f(\underline{u}) &= f(\underline{u} + d\underline{u}) - f(\underline{u}) \\
&= (1 + d) \cdot f(\underline{u}) - f(\underline{u}) \\
&= d \cdot f(\underline{u})
\end{aligned}
$$

and thus

$$
f(\underline{u}) = \partial_{\underline{u}} f(\underline{u}) = \partial_{\underline{u}} f(\underline{0}) = df(\underline{0})(\underline{u}).
$$

which, according to Proposition 7, is linear.   ∎

We can generalize this fact to other degrees of homogeneity.

Definition 1. *Let $f : V \to E$. We say that f is k-homogenous if*

$$
\forall \lambda \in R, \ \forall \underline{v} \in V, \ f(\lambda \cdot \underline{v}) = \lambda^k f(\underline{v}),
$$

and, f is said to be of degree k if there exists a k-linear mapping $\varphi : V^k \to E$ such that

$$
\forall \underline{v} \in V, \ f(\underline{v}) = \varphi(\underline{v}, \ldots, \underline{v}).
$$

Proposition 11. *The following propositions are equivalent:*

1. *f is k-homogeneous;*

2. *$\forall \underline{v} \in V \ (\partial_{\underline{v}} \cdots \partial_{\underline{v}} f)(0) = k! \, f(\underline{v})$;*

3. *f is of degree k.*

Proof: Obviously (3) implies (1) and it is clear that (2) implies (3) if one lets

$$
\varphi(\underline{v_1}, \cdots, \underline{v_k}) = \frac{1}{k!}(\partial_{\underline{v_1}}, \cdots, \partial_{\underline{v_k}} f)(0).
$$

We now prove that (1) implies (2).

If $g : V \to E$, we consider, for fixed $\underline{v}$, the function $R \to E$: $\lambda \to g(\lambda.\underline{v})$. Differentiating with respect to λ we have

$$
\begin{aligned}
d \cdot \frac{\partial}{\partial \lambda} g(\lambda \cdot \underline{v}) &= g((\lambda + d)\underline{v}) - g(\lambda \underline{v}) \\
&= d \cdot (\partial_{\underline{v}} g)(\lambda \underline{v})
\end{aligned}
$$

and, thus, $\frac{\partial}{\partial\lambda}g(\lambda\cdot\underline{v})=(\partial_{\underline{v}}g)(\lambda\cdot\underline{v})$. Iterating we find (for $1\leq r\leq k$)

$$\frac{\partial^r}{\partial\lambda^r}g(\lambda\cdot\underline{v})=(\partial_{\underline{v}}\cdots\partial_{\underline{v}}g)(\lambda\cdot\underline{v}).$$

Coming back to k-homogenous f, we find

$$
\begin{aligned}
(\partial_{\underline{v}}\cdots\partial_{\underline{v}}f)(\lambda\cdot\underline{v}) &= \frac{\partial^k}{\partial\lambda^k}f(\lambda\underline{v}) \\
&= \frac{\partial^k}{\partial\lambda^k}(\lambda^k\cdot f(\underline{v})) \\
&= k!\cdot f(\underline{v}).
\end{aligned}
$$

Letting $\lambda=0$, we have (2).   ■

1.3 The rudimentary integral calculus

1.3.1 Preorder

We now introduce a preorder relation on our ring R modelling the physical straight line, . More precisely, we fix a relation on R, denoted $x\leq y$, satisfying the following conditions :

Preorder axiom.

1. The relation $\leq$ is a preorder, i.e. a reflexive and transitive relation ;

2. This preorder is compatible with the unitary ring structure, i.e.

    (a) $\forall x,y,z\in R\ (x\leq y)\rightarrow(x+z\leq y+z)$,
    (b) $\forall x,y,\in R\ (0\leq x\ and\ 0\leq y)\rightarrow(0\leq x.y)$,
    (c) $0\leq 1\ and\ not\ (1\leq 0)$

3. $\forall d\in D\ 0\leq d\ and\ d\leq 0.$

The last condition shows the reason why we have not insisted that the relation $\leq$ be an order. Requiring anti-symmetry would entail $D=\{0\}$.

We define
$$[a, b] = \{x \in R \mid a \leq x \leq b\}.$$
Condition (3) tells us that

$$D \subseteq [0, 0].$$

Also note that for all d_1 of D and all d_2 of D one has

$$[a, b] = [a + d_1, b + d_2].$$

This last remark guarantees us that the elementary differential calculus extends from functions defined on R to functions defined on $[a, b]$. For instance, if $f : [a, b] \rightarrow R$ and $x \in [a, b]$, we can define $f'(x)$ by the formula
$$\forall d \in D \; f(x + d) = f(x) + d \cdot f'(x)$$
since $x \in [a, b] \rightarrow x + d \in [a, b]$. Thus if f is defined on $[a, b]$, so will f' be.

We need the following elementary property :

Proposition 1. *Each interval $[a, b]$ is convex. In other words if $x, y \in [a, b]$ and $\lambda \in [0, 1]$, then $x + \lambda(y - x) \in [a, b]$.*

Proof: From $y \leq b$ and $0 \leq \lambda$ we deduce $\lambda(y - x) \leq \lambda(b - x)$. But from $x \leq b$ and $\lambda \leq 1$ we also deduce $\lambda(b - x) \leq b - x$. Whence $x + \lambda(y - x) \leq b$. In the same way $a \leq x + \lambda(y - x)$. ∎

As usual, $x < y$ is defined by ($x \leq y$ *and* $x \neq y$). For instance, we have $0 < 1$, since we postulated $0 \neq 1$ and $0 \leq 1$. We put

$$]a, b[= \{x \in R \mid a < x < b\}.$$

We immediately have
$$]a, b[\subseteq [a, b].$$

A remark is in order : as this deals with axiomatics, we cannot add axioms without first convincing ourselves that not only models exist but also that there are useful models for classical differential geometry. For instance, it is not reasonable to require that the preorder be antisymmetric or total ($x \leq y$ *or* $y \leq x$).

On the other hand (cfr. [4] or [5]), there are interesting models satisfying the axiom

$$x \; invertible \rightarrow (x \geq 0 \; or \; x \leq 0).$$

A. Kock (cf. [4]) showed that one can accept the axiom

$$x \; invertible \leftrightarrow not \; (x = 0).$$

Note that, from this axiom and from the fact that $d \in D$ is not invertible we can only deduce *not (not $d = 0$)*, which does not imply $d = 0$ in intuitionistic logic.

As a simple exercise (which will not be used afterwards) the reader may prove the following proposition.

Proposition 2. *If the following axioms hold*

$$not \; (x = 0) \rightarrow x \; invertible \rightarrow (x \geq 0 \; or \; x \leq 0)$$

then

$$(x \geq 0 \; and \; not \; (x = 0)) \leftrightarrow not \; (x \leq 0).$$

1.3.2   Integral on the interval $[0, 1]$

A. Kock and G. Reyes [10] proposed the following integration axiom :

(Integration axiom). **For every function $f : [0,1] \rightarrow R$, there exists a unique function $g : [0, 1] \rightarrow R$ such that $g' = f$ and $g(0) = 0$.**

For each x in $[0, 1]$, the value $g(x)$ of the unique function g such that $g' = f$ and $g(0) = 0$ will be denoted $\int_0^x f(t)dt$ or $\int_0^x f$.

We have the following elementary properties :

Proposition 3. *For every f and g from $[0, 1]$ to R and every α in R,*

1. $\int_0^x (f + g) = \int_0^x f + \int_0^x g$;

2. $\int_0^x \alpha \cdot f = \alpha \cdot \int_0^x f$;

3. $\int_0^x f' = f(x) - f(0)$;

4. $\int_0^x f' \cdot g = (f(x)g(x) - f(0)g(0)) - \int_0^x f \cdot g'$.

Proof: It suffices to observe that in each case the two sides of the equality have the same value zero at 0 and that they have the same derivative. ∎

An essential step in the extension of the notion of integral to other intervals is the following proposition, known as "Hadamard's lemma" .

Proposition 4. Let $a, b \in R$ and let $f : [a, b] \to R$. Then , for any x and y in $[a, b]$,

$$f(y) - f(x) = (y - x) \int_0^1 f'(x + t(y - x))dt.$$

Proof: According to Proposition 1, the function $\varphi(t) = x + t(y - x)$ defined on $[0, 1]$ does take its values in $[a, b]$ since x and y are in $[a, b]$. Since $\varphi'(t) = (y - x)$, the right hand side is

$$\int_0^1 f'(\varphi(t)) \cdot \varphi'(t)dt = \int_0^1 (f \circ \varphi)' = f(\varphi(1)) - f(\varphi(0)) = f(y) - f(x).$$

∎

Let us point out the rule of differentiation under the integral sign with respect to a parameter.

Proposition 5. Let $f : [0, 1] \times R \to R$. Put $g(s) = \int_0^1 f(t, s)dt$. One has

$$g'(s) = \int_0^1 \frac{\partial f}{\partial s}(t, s)dt.$$

Proof: For every d in D

$$
\begin{aligned}
g(s + d) - g(s) &= \int_0^1 [f(t, s + d) - f(t, s)]dt \\
&= d \cdot \int_0^1 \frac{\partial f}{\partial s}(t, s)dt,
\end{aligned}
$$

hence the claim follows from the definition of the derivative. ∎

1.3.3   Integral on an interval [a, b]

Let $a \leq b$ in R and $f : [a, b] \to R$. We intend to show the existence of a unique function $g : [a, b] \to R$ such that $g' = f$ and $g(a) = 0$. Let us begin with the following uniqueness result.

Proposition 6. *Let $f : [a, b] \to R$. If $f' = 0$, then f is constant.*

Proof: By Hadamard's lemma (Proposition 4), for all x in $[a, b]$

$$
\begin{aligned}
f(x) - f(a) &= (x - a) \int_0^1 f'(a + t(x - a))dt \\
&= (x - a) \int_0^1 0 = 0.
\end{aligned}
$$

∎

Proposition 7. *Let $f : [a, b] \to R$. Then, there exists a unique function $g : [a, b] \to R$ such that $g' = f$ and $g(a) = 0$.*

Proof: The uniqueness being a result of the previous proposition, we prove the existence. This is done explicitly by defining :

$$ g(x) = (x - a) \int_0^1 f(a + t(x - a))dt. $$

We have $g(a) = 0$ and

$$ g'(x) = \int_0^1 f(a + t(x - a))dt + (x - a)\frac{d}{dx} \int_0^1 f(a + t(x - a))dt $$

which, according to Proposition 5, gives

$$
\begin{aligned}
g'(x) &= \int_0^1 f(a + t(x - a))dt + (x - a) \int_0^1 \frac{\partial}{\partial x} f(a + t(x - a))dt \\
&= \int_0^1 f(a + t(x - a))dt + (x - a) \int_0^1 t \cdot f'(a + t(x - a))dt
\end{aligned}
$$

Set $\varphi(t) = a + t(x - a)$. We have $\varphi'(t) = x - a$ so we can write

$$
\begin{aligned}
g'(x) &= \int_0^1 f(\varphi(t))dt + \int_0^1 t \cdot f'(\varphi(t)) \cdot \varphi'(t) \cdot dt \\
&= \int_0^1 f(\varphi(t))dt + \int_0^1 t \cdot (f \circ \varphi)'(t)dt
\end{aligned}
$$

whence, computing the second integral by parts :

$$= \int_0^1 f(\varphi(t))dt + [t \cdot f(\varphi(t))]_0^1 - \int_0^1 f(\varphi(t))dt$$
$$= f(\varphi(1))$$
$$= f(x),$$

which finishes the proof. ∎

For all x in $[a, b]$, the value $g(x)$ of the unique function g such that $g' = f$ and $g(a) = 0$, will be denoted by

$$\int_a^x f(t)dt \ \ or \ \ \int_a^x f.$$

Since this integral is expressed in terms of an integral on $[0, 1]$, Proposition 3 is immediately extended.

Proposition 8. *For all f and g from $[a, b]$ to R and all α in R :*

a) $\int_a^b (f + g) = \int_a^b f + \int_a^b g$;

b) $\int_a^b \alpha \cdot f = \alpha \cdot \int_a^b f$

c) $\int_a^b f' = f(b) - f(a)$

d) $\int_a^b f' \cdot g = [f(x) \cdot g(x)]_a^b - \int_a^b f \cdot g'.$

∎

Let us prove the formula of integration by substitution :

Proposition 9. *Let $a \leq b$ and $c \leq d, \varphi : [a, b] \to [c, d]$ with $\varphi(a) = c$, $\varphi(b) = d$ and let $f : [c, d] \to R$. We have*

$$\int_c^d f(s)ds = \int_a^b f(\varphi(t)) \cdot \varphi'(t)dt.$$

Proof: The two functions from $[a, b]$ to R given by

$$F_1(u) = \int_c^{\varphi(u)} f(s)ds$$
$$F_2(u) = \int_a^u f(\varphi(t))\varphi'(t)dt$$

vanish for $u = a$ and have the same derivative for all u in $[a, b]$. By Proposition 6, they coincide on $[a, b]$. ∎

The following proposition states the additivity of the integral.

Proposition 10. *Let* $a \leq c \leq b$ *and* $f : [a, b] \to R$. *We have*

$$\int_a^b f(t)dt = \int_a^c f(t)dt + \int_c^b f(t)dt.$$

Proof: Consider the function

$$F : [a, b] \to R$$

defined by

$$F(x) = \int_a^x f(t)dt + \int_x^b f(t)dt$$

We have $F(a) = \int_a^b f(t)dt$. To prove the result, let us prove that F is constant by showing that its derivative is zero . The derivative of $\int_a^x f(t)dt$ is $f(x)$. Let us compute the derivative of $\int_x^b f(t)dt$. Consider the change of variable

$$[0, b - x] \to [x, b] : u \rightsquigarrow t = x + u.$$

According to the previous proposition we have

$$\int_x^b f(t)dt = \int_0^{b-x} f(x + u)du.$$

The derivative with respect to x is then

$$(-1).f(x+(b-x))+\int_0^{b-x} f'(x+u)du = -f(b)+[f(x+u)]_0^{b-x} = -f(x).$$

Thus $F'(x) = 0$, which proves the proposition.   ∎

Let us point out the infinitesimal mean value theorem, which is just the following formula :

Proposition 11. $\int_a^{a+d} f(t)dt = d \cdot f(a)$ *for all* d *in* D.

Proof: Since $\frac{d}{dx}(\int_a^x f(t)dt) = f(x)$, we have by definition of the derivative

$$\int_a^{x+d} f - \int_a^x f = d \cdot f(x)$$

and thus the thesis follows by putting $x = a$.   ∎

One can then pass easily to the case of integrals of functions with n variables defined on an n-interval of R^n. We shall only prove Fubini's theorem :

Proposition 12. *Let $f : [a_1, b_1] \times [a_2, b_2] \to R$. We have :*

$$\int_{a_1}^{b_1} (\int_{a_2}^{b_2} f(t,s)ds)dt = \int_{a_2}^{b_2} (\int_{a_2}^{b_1} f(t,s)dt)ds.$$

Proof: Let

$$\psi(x) = \int_{a_2}^{b_2} (\int_{a_1}^{x} f(t,s)dt)ds$$

for $x \in [a_1, b_1]$. The righthand side of the equality to be proved is $\psi(b_1)$. On the one hand, we have $\psi(a_1) = 0$ and, considering Proposition 5,

$$\begin{aligned} \psi'(x) &= \int_{a_2}^{b_2} \frac{\partial}{\partial x} (\int_{a_1}^{x} f(t,s)dt)ds \\ &= \int_{a_2}^{b_2} f(x,s)ds. \end{aligned}$$

Thus ψ being the only function on $[a_1, b_1]$ vanishing at a_1 and whose derivative has the indicated value, we have :

$$\psi(b_1) = \int_{a_1}^{b_1} (\int_{a_2}^{b_2} f(t,s)ds)dt$$

as announced. ∎

1.3.4   A few applications of the axiom of integration

The axiom of integration is more powerful than it seems at first sight. We intend to show some of its consequences.

The first and easy application is Taylor's formula with remainder :

Proposition 13. *Let $f : [a, b] \to R$ and $x \in [a, b]$. We have for each n :*

$$f(x) = \sum_{k=0}^{k} \frac{f^{(k)}(a)}{k!} (x-a)^k + \int_{a}^{x} f^{(n+1)}(t) \cdot \frac{(x-t)^n}{n!} dt.$$

Proof: We proceed by induction on n. For $n = 0$, the formula is

$$f(x) = f(a) + \int_a^x f'(t)dt$$

which is part (c) of Proposition 8. Consider

$$I_n = \int_a^x f^{(n+1)}(t) \cdot \frac{(x-t)^n}{n!} dt$$

and compute it by parts. We find

$$
\begin{aligned}
I_n &= [-\frac{(x-t)^{n+1}}{(n+1)!} f^{(n+1)}(t)]_a^x + I_{n+1} \\
&= \frac{(x-a)^{n+1}}{(n+1)!} f^{(n+1)}(a) + I_{n+1}
\end{aligned}
$$

which finishes the proof.   ∎

Furthermore, note that the integration axiom extends not only to the interval $[a,b]$ (Proposition 7) but also to the whole of R :

Proposition 14. *Let $f : R \to R$ and $x_0 \in R$. There is a unique function $g : R \to R$ such that $g' = f$ and $g(x_0) = 0$.*

Proof: As in Proposition 7, the existence can be proved explicitly by putting

$$g(x) = (x - x_0) \int_0^1 f(x_0 + t(x - x_0))dt.$$

As we have, of course, Hadamard's lemma (Proposition 4) for functions from R to R, the uniqueness follows as in Proposition 6.   ∎

Just as in the classical theory, it follows from this proposition that two functions having the same derivative differ only by a constant.

A crucial point for what follows is :

Proposition 15. *Let $f : R \to R$. If $\forall x \in R, x \cdot f(x) = 0$, then $\forall x \in R$, $f(x) = 0$.*

Proof: Suppose that for all x in R, $x \cdot f(x) = 0$ and let $\lambda \in R$. Let us show that $f(\lambda) = 0$. Define $\varphi_\lambda : R \to R$ by

$$\varphi_\lambda(x) = x \cdot f(x \cdot \lambda).$$

Let us compute the derivative of φ_λ :

$$(\varphi_\lambda)'(x) = f(x \cdot \lambda) + x \cdot \lambda \cdot f'(x \cdot \lambda).$$

Since, by assumption, $\varphi_1(x) = x \cdot f(x)$ is the zero function, we have for all u in R

$$0 = \varphi_1'(u) = f(u) + u \cdot f'(u).$$

Hence $(\varphi_\lambda)'(x) = (\varphi_1')(x \cdot \lambda) = 0$. As on the other hand $\varphi_\lambda(0) = 0$, φ_λ is the zero function. In particular its value at $x = 1$ is zero, i.e. $f(\lambda) = 0$.
∎

G. Reyes proposed to use as a basic axiom for differential calculus the following condition : for each function $f : R \to R$ there exists a unique function $g : R \times R \to R$ such that

$$\forall x, y \in R \ f(x) - f(y) = (x - y)g(x, y).$$

G. Reyes proposed to call that axiom "Fermat's axiom" because Fermat observed that, if f is a polynomial, then $f(x) - f(a)$ is divisible by $x - a$. More generally, one can consider the quotient $g(x, a)$, and Fermat used $g(a, a)$ for the study of tangents ; this of course anticipates the notion of derivative.

Having assumed the Kock-Lawvere axiom and integration axiom, the Fermat-Reyes axiom is in fact a theorem :

Proposition 16. *For all $f : R \to R$, there exists a unique function $g : R \times R \to R$ such that $\forall x, y \in R$*

$$f(x) - f(y) = (x - y)g(x, y).$$

Proof: The existence of g follows from Hadamard's lemma (Proposition 4). The uniqueness follows from the previous proposition. ∎

The unique function g of this proposition will be denoted by ∂f. It contains all the information concerning the derivative of f.

Proposition 17. *We have* $f'(x) = \partial f(x, x)$.

Proof: By definition of ∂f, we have

$$f(x + d) - f(x) = d \cdot \partial f(x + d, x).$$

But for every function $h : D \to R$ we have

$$d \cdot h(d) = d \cdot (h(0) + d \cdot h'(0)) = d \cdot h(0).$$

Thus

$$f(x + d) - f(x) = d \cdot \partial f(x, x).$$

Since this holds for every d in D, we have the desired result. ∎

Remember that an element x in R is invertible if

$$\exists\, y (xy = 1).$$

We denote this formula by $Inv(x)$. Of course, if the inverse y of x exists, it is unique; it will be denoted by $1/x$.

An interesting extension of Fermat's property is l'Hôpital's rule which will be stated as follows.

Proposition 18. *Let* $f, g : R \to R$ *with* $f(0) = g(0) = 0$. *If for all* x *of* R *we have* $Inv(\partial g(x, 0))$, *then there exists a unique function* $h : R \to R$ *such that*

$$\forall\, x \in R \; h(x) \cdot g(x) = f(x).$$

Furthermore, this function satisfies :

$$h(0) \cdot g'(0) = f'(0).$$

Proof: Since $f(0) = g(0) = 0$, we have

$$
\begin{aligned}
f(x) &= x \cdot \partial f(x, 0) \\
g(x) &= x \cdot \partial g(x, 0)
\end{aligned}
$$

The existence of the function h is proved explicitly by putting

$$h(x) = \partial f(x, 0) / \partial g(x, 0)$$

because

$$h(x) \cdot g(x) = (\partial f(x,0)/\partial g(x,0)) \cdot x \cdot \partial g(x,0)$$
$$= x \cdot \partial f(x,0)$$
$$= f(x)$$

and furthermore we have

$$h(0) = \partial f(0,0)/\partial g(0,0) = f'(0)/g'(0)$$

according to Proposition 17.

Let us check uniqueness. It suffices of course to see that if

$$\forall\, x \in R, \ h(x) \cdot g(x) = 0$$

then $\forall\, x \in R$, $h(x) = 0$. But since $g(x) = x \cdot \partial g(x,0)$, we have $h(x) \cdot x \cdot \partial g(x,0) = 0$ and since $\partial g(x,0)$ is invertible, $x \cdot h(x) = 0$. Uniqueness then follows from Proposition 15. ∎

The previous arguments can be iterated to get Taylor's formulas and l'Hôpital's rules of arbitrary degree k.

Proposition 19. *Let k be a given positive integer. For all $f : R \to R$, there exists a unique function $g_k : R \times R \to R$ such that $\forall\, x, y \in R$*

$$f(y) - f(x) = (y - x)f'(x) + (y - x)^2\frac{f''(x)}{2} + \ldots$$
$$+ \ldots + (y - x)^{k-1}\frac{f^{(k-1)}(x)}{(k-1)!} + (y - x)^k g_k(y, x)$$

and this function g_k satisfies

$$g_k(x, x) = \frac{f^{(k)}(x)}{k!}.$$

Proof:

a) Uniqueness is immediate : it suffices to apply k times Proposition 15.

b) Let us prove the existence clause by using induction on k. For $k = 1$, it is just Proposition 16. Assume the existence of a function $h : R \times R \to R$ such that

$$f(y) - f(x) = \sum_{i=1}^{k-2} (y - x)^i \frac{f^{(i)}(x)}{i!} + (y - x)^{k-1} h(y, x)$$

with $h(x, x) = \frac{f^{(k-1)}(x)}{(k-1)!}$. For each x let us apply Proposition 16 to the function $h(\cdot, x)$. There exists a function ∂h such that

$$h(y_2, x) - h(y_1, x) = (y_2 - y_1)\partial h(y_2, y_1, x).$$

Choosing $y_1 = x$ and putting $g_k(y, x) = \partial h(y, x, x)$, we find

$$h(y, x) - \frac{f^{(k-1)}(x)}{(k-1)!} = (y - x)g_k(y, x)$$

whence the existence of the function g_k we sought for.

c) It remains to show that $g_k(x, x) = \frac{f^{(k)}(x)}{k!}$. Let $d_1, \ldots, d_k \in D$. Let us put $\delta = d_1 + \ldots + d_k$ and $y = x + \delta$. We have

$$f(x + \delta) - f(x) = \sum_{i=1}^{k-1} \delta^i \frac{f^{(i)}(x)}{i!} + \delta^k g_k(x + \delta, x).$$

Now, as $\delta = d_1 + \ldots + d_k$,

$$\delta^k g_k(x + \delta, x) = \delta^k g_k(x, x).$$

But then Proposition 2.6 (uniqueness of Taylor development) gives us

$$g_k(x, x) = \frac{f^{(k)}(x)}{k!}$$

as we wanted.

∎

Instead of $g_k(y, x)$ we shall write $\partial_k f(y, x)$ in order to point out the dependence on f. We then have l'Hospital's rule of order k :

Proposition 20. *Let* $f, g : R \rightarrow R$ *with* $f(0) = f'(0) = \ldots = f^{(k-1)}(0) = 0$ *and* $g(0) = g'(0) = \ldots = g^{(k-1)}(0) = 0$. *If for all* x *of* R *we have* $Inv(\partial_k g(x, 0))$, *then there exists a unique function* $h : R \rightarrow R$ *such that*

$$\forall \, x \in R, \; h(x) \cdot g(x) = f(x).$$

Furthermore we have

$$h(0) \cdot g^{(k)}(0) = f^{(k)}(0).$$

Proof: From the previous proposition we have

$$
\begin{aligned}
f(x) &= x^k \partial_k f(x, 0) \\
g(x) &= x^k \partial_k g(x, 0).
\end{aligned}
$$

The function h is then given explicitly by :

$$h(x) = \frac{\partial_k f(x, 0)}{\partial_k g(x, 0)}$$

and we do find

$$h(0) = \frac{\partial_k f(0, 0)}{\partial_k g(0, 0)} = \frac{f^{(k)}(0)}{g^{(k)}(0)}.$$

Let us verify the uniqueness of h. To this end suppose that $\forall \, x \in R$, $h(x)g(x) = 0$. Then we have

$$h(x) \cdot x^k \cdot \partial_k g(x, 0) = 0$$

and thus $x^k h(x) = 0$ and it suffices to apply k times Proposition 15 to get $h(x) = 0$. ∎

1.4   Commented bibliography

1) The prehistory of synthetic differential geometry can be found in the work of algebraist geometers. André Weil can be considered as the first to make a start [56](1953). Weil himself says that his theory has a double source, on the one hand the return to Fermat's methods in first

order infinitesimal calculus and, on the other hand, the theory of jets developed by Charles Ehresmann (cf [17]).

Another fundamental contributor is Alexandre Grothendieck [22] (1960-1967) who insisted on the systematic inclusion of nilpotents in rings considered for algebraic geometry. From the very first section of his treatise, he was quite inflexible on the point.

2) Strictly speaking, the starting point of the theory rests in the conferences of F. William Lawvere [44] in 1967. These conferences were published in [25] (1979), edited by Anders Kock.

3) The first book on synthetic differential geometry is Kock's book : "Synthetic Differential Geometry" [31] (1981). Another book appeared in 1991, being the work of Ieke Moerdijk and Gonzalo E. Reyes [52].

We must also mention the conferences held at Aarhus in 1978 and 1983. They led to two miscellaneous collections edited by Kock([25] (1979) and [26] (1983)).

4) Concerning the content of this first chapter, we note that the starting point - the Lawvere axiom and the categorical bases of the analysis - is in Lawvere's lectures. They provide the ground ideas for synthetic differential calculus, which was first developed by Kock. The main part of our Paragraph 1.2 appears for the first time in [27] (1977) and [28] (1978).

Lawvere had already introduced an integration axiom in [44]. Kock and Reyes developed this idea [38] (1981). Belair's thesis [2] (1981), prepared under the supervision of Reyes, gives some basic differential and integral calculus ; it also discusses other axioms, such as, for instance, a positivity axiom of the integral, an axiom of the intermediate tranversal value, and an axiom of the existence of trigonometrical functions. Other suggestions of axioms on the existence of local solutions of differential equations are explored in C. McLarty [46] (1983) and still others in papers of Marta Bunge and Eduardo Dubuc ([7], [8]).

5) An important step in the development of synthetic analysis was achieved with Jacques Penon's thesis [53], which analyses the problem of the local inverse theorem. This has opened the way to other existence problems considered by Bunge, Dubuc, F.Gago and others in [6], [8], [9], [19], [20]. We do not develop these results in the present book.

Chapter 2

Weil algebras and infinitesimal linearity

2.1 Weil algebras

2.1.1 Introduction

We encountered in 1.1 various small objects, the most typical and simple one being D. We intend to indicate in this section a more algebraic view of these small objects.

As an example, let us examine what happens in the simple case of D.

Let $R[X]$ be the ring of polynomials in X with coefficients in R. In this R-algebra, let I be the ideal generated by the polynomial X^2 and put

$$W = R[X]/(X^2).$$

By construction, W is a finitely-presented R-algebra, i.e. the quotient of a free R-algebra of a finite type (in this case with only one generator X) by an ideal of finite type (in this case with a single generator X^2).

Let us now consider the *spectrum* of W : it associates to an R-algebra C the set

$$spec_C(W) = \{c \in C \mid c^2 = 0\}$$

of elements in C annihilating the polynomials of I.

Thus
$$D = spec_R(W).$$

The R-algebra W possesses another description. Denoting by ϵ the equivalence class of X modulo I, one has $\epsilon^2 = 0$ and W can be described as $\{a_0 + a_1\epsilon \mid a_0, a_1 \in R\}$ equipped with the product

$$(a_0 + a_1\epsilon)(b_0 + b_1\epsilon) = a_0b_0 + (a_0b_1 + a_1b_0)\epsilon.$$

As an R-module, W is just R^2. Note that we have a homomorphism

$$W \to R : (a_0 + a_1\epsilon) \rightsquigarrow a_0$$

whose kernel is an ideal $J = \{a_1\epsilon \mid a_1 \in R\}$ with null square .

We have a homomorphism of R-algebras

$$W \xrightarrow{\alpha} R^D$$

that associates to $a_0 + a_1\epsilon$ the function of D into R given by $f(d) = a_0 + a_1d$. It is now possible to impose the Kock-Lawvere axiom by asking that α be an isomorphism of R-algebras.

In this section we intend to describe a class of algebras called Weil algebras, corresponding to other small objects, and to state in terms of these small objects an axiom generalizing (K-L).

2.1.2   Weil algebras

An augmented commutative unitary R-algebra of finite dimension consists of a finite dimensional free module $R^{n+1} = \{(a_0, a_1, \ldots, a_n) \mid a_i \in R\}$ equipped with a bilinear, associative, commutative multiplication

$$m : R^{n+1} \times R^{n+1} \to R^{n+1},$$

having $(1, 0, \ldots, 0)$ as unit element and such that

$$R^{n+1} \xrightarrow{\pi_0} R : (a_0, a_1, \ldots, a_n) \rightsquigarrow a_0$$

is a homomorphism. We put $J = Ker\ \pi_0$ and call it the ideal of augmentation. As an R-module, J is isomorphic to R^n :

$$J = \{(0, a_1, \ldots, a_n) \mid a_i \in R\}.$$

Note that, R not being a field, it can have non trivial ideals and thus J is not necessarily a maximal ideal. So, we do not, in general, have a local ring.

Definition 1. *A Weil-algebra W is an augmented unitary commutative R-algebra of finite dimension whose ideal of augmentation J is nilpotent. A homomorphism of Weil-algebras*

$$f : W_1 \to W_2$$

is a homomorphism of R-algebras sending the ideal of augmentation J_1 from W_1 into the ideal of augmentation J_2 of W_2.

This amounts to asking that the diagram

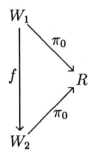

commutes. In fact if R is a field, a homomorphism of Weil-algebras is simply a homomorphism of R-algebras because the homomorphic character of f implies that the image of a nilpotent element is nilpotent and if R is a field, then the ideal of augmentation can only be the set of nilpotent elements ; if R is a field, then W is a local ring.

We now indicate what is a finite presentation of a Weil-algebra. Let $e_0, e_1, \ldots, e_n$ be the canonical basis of R^{n+1}. Of course $e_0 = (1, 0, \ldots, 0)$ is the unit element. The bilinear multiplication is thus determined by the structure constants expressing the products $e_i \cdot e_j$ (for $1 \leq i, j \leq n$) :

$$e_i \cdot e_j = \sum_{k=1}^{n} \gamma_{ij}^k e_k.$$

Let us consider the free algebra with n-generators

$$R[X_1, X_2, \ldots, X_n]$$

and the homomorphism of unitary R-algebras

$$R[X_1, X_2, \ldots, X_n] \xrightarrow{\pi} W$$

defined by $\pi(X_i) = e_i$. The kernel of π is the ideal I generated by the polynomials

$$X_i X_j - \sum_{k=1}^{n} \gamma_{ij}^k X_k \tag{1}$$

for $1 \leq i, j \leq n$. Thus, modulo I, a polynomial P is a polynomial of first degree (as I contains (1)) and a polynomial of first degree is in $Ker \, \pi$ if and only if it is null, as $(e_0, \ldots, e_n)$ is a basis of W.

Note that the elements e_i ($1 \leq i \leq n$) being nilpotent, there exists an integer m large enough for the X_i^m to be in I. The minimum number of generators of W, as an R-algebra, is called the *breadth* of W. Note also that the ideal of augmentation J being nilpotent, there exists a least integer m such that $J^{m+1} = (0)$; this m is called the *height* of W.

Let us consider a presentation of the Weil-algebra W :

$$W \simeq R[X_1, X_2, \ldots, X_r]/I$$

where I is generated by the polynomials $P_1, P_2, \ldots, P_s$. Let C be an R-algebra. We define the *spectrum* of W in C as :

$$Spec_C(W) = \{(a_1, \ldots, a_r) \in C^r \mid P_j(\bar{a}) = 0, \; j = 1, 2, \ldots, s\}.$$

In particular, we put

$$D(W) = Spec_R(W) \subseteq R^r.$$

Such a spectrum of W in R is called a *small object*.

So we find for each R-algebra C a "contravariant functor" from the category of Weil-algebras to the one of "sets" (of our intuitionist theory). Indeed, let us consider a homomorphism

$$f : W_1 \to W_2$$

and presentations

$$\begin{aligned} W_1 &\simeq R[X_1, \ldots, X_r]/(P_1, P_2, \ldots, P_s) \\ W_2 &\simeq R[Y_1, \ldots, Y_{r'}]/(Q_1, Q_2, \ldots, Q_{s'}), \end{aligned}$$

Let C be an R-algebra. We define

$$Spec_C(f) : Spec_C(W_2) \to Spec_C(W_1)$$

as follows. Given

$$(b_1, b_2, \ldots, b_{r'}) \in Spec_C(W_2),$$

define

$$Spec_C(f)(b_1, b_2, \ldots, b'_r) = (a_1, a_2, \ldots, a_r)$$

where

$$a_i = f(X_i)\,(b_1, b_2, \ldots, b'_r),$$

which makes sense because, although $f(X_i)$ is defined only modulo $(Q_1, Q_2, \ldots, Q_s)$, its value at $\underline{b}$ is still determined, as the Q_j are zero at $\underline{b}$. The functoriality of this construction is immediate (i.e.

$$Spec_C(id_W) = id_{Spec_C(W)}, \ Spec_C(g \circ f) = Spec_C(f) \circ Spec_C(g)).$$

Let us indicate now a few examples of Weil-algebras.

1) The Weil-algebras of breadth 1 are the algebras

$$W_k = R[X]/X^{k+1}.$$

By denoting ϵ the equivalence class of X into W_k, we see that we can describe W_k as an algebra

$$\{a_0 + a_1\epsilon + \ldots + \alpha_k\epsilon^k \mid a_i \in R\} \simeq R^{k+1}$$

equipped with the usual multiplication of polynomials, considering that $\epsilon^{k+1} = 0$. Thus we see that W_k is of height k.

We have

$$D(W_k) = \{\delta \in R \mid \delta^{k+1} = 0\}.$$

We denote this object by D_k.

2) Let us recall that we defined (1.1.3) the objects

$$D(2) = \{(d_1, d_2) \mid d_1^2 = d_2^2 = d_1d_2 = 0\}$$

and more generally

$$D(n) = \{(d_1, d_2, \ldots, d_n) \mid d_id_j = 0 \text{ for } 1 \le i, j \le n\}.$$

They are associated to the Weil algebras

$$W(n) = R[X_1, X_2, \ldots, X_n]/_{(\{X_i X_j | 1 \leq i, j \leq n\})}.$$

Denoting by ϵ_i the equivalence class of X_i in $W(n)$, we see that one can indentify $W(n)$ with the algebra

$$\{a_0 + a_1 \epsilon_1 + a_2 \epsilon_2 + \ldots + a_n \epsilon_n \mid a_i \in R\} \simeq R^{n+1}$$

equipped with the multiplication taking into account the identities $\epsilon_i \epsilon_j = 0$. We see that $W(n)$ is of height 1.

Obviously we have $D(W(n)) = D(n)$.

3) Let us observe that $D \times D$ is the spectrum in R of the Weil-algebra $W_1 \otimes W_1$ that can be described as

$$R[X, Y]/(X^2, Y^2)$$

which shows that it is of breadth 2. One can also describe it as

$$\{a_0 + a_1 \epsilon_1 + a_2 \epsilon_2 + a_3 \epsilon_1 \epsilon_2 \mid a_i \in R\}$$

where ϵ_1 (resp. ϵ_2) is the class of X (resp. Y). Thus it is a Weil-algebra of height 2.

More generally, let W_1 and W_2 be two Weil-algebras. Let us consider presentations of W_1 and W_2

$$\begin{aligned} W_1 &= R[X_1, \ldots, X_r]/(P_1, \ldots, P_s) \\ W_2 &= R[Y_1, \ldots, Y_u]/(Q_1, \ldots, Q_v) \end{aligned}$$

The sum of these two Weil-algebras, which is also their tensor product as R-algebras, can be presented by :

$$W_1 \otimes W_2 = R[X_1, \ldots, X_r, Y_1, \ldots, Y_u]/(P_1, \ldots, P_s, Q_1, \ldots, Q_v)$$

and is still a Weil-algebra.

Notice that

$$\begin{aligned} D(W_1 \otimes W_2) = \{(a_1, \ldots, a_r, b_1, \ldots, b_u \mid P_1(\underline{a}) = \ldots = P_s(\underline{a}) = \\ = Q_1(\underline{b}) = \ldots = Q_v(\underline{b}) = 0\} \\ \simeq D(W_1) \times D(W_2). \end{aligned}$$

The product of two small objects is thus small.

4) As an easy exercise, the reader could verify, for instance, that

$$\{(d_1, d_2, d_3)|d_1^2 = d_2^2 = d_3^2 = d_1 \cdot d_3 = d_2 \cdot d_3 = 0\}$$

or

$$\{(d_1, d_2, e_1, e_2)|d_i^2 = e_i^2 = d_i \cdot e_j = e_1 \cdot e_2 = 0\}$$

are small objects.

5) Let us consider the R-algebra presented by

$$W = R[X_1, X_2, \ldots, X_n, Y_1, Y_2, \ldots, Y_n]/I$$

where I is generated by

$$\{X_i X_j \mid 1 \leq i, j \leq n\} \ \cup \ \{Y_i Y_j \mid 1 \leq i, j \leq n\}$$
$$\cup \ \{X_i Y_j + X_j Y_i \mid 1 \leq i \leq j \leq n\}.$$

Let ϵ_i denote the class of X_i modulo I, and η_i that of Y_i (for $1 \leq i \leq n$). We show that, as an R-module, W has a base comprising : 1, the ϵ_i $(1 \leq i \leq n)$, the η_i $(1 \leq i \leq n)$ and the elements $\epsilon_i \eta_j - \epsilon_j \eta_i$ for $1 \leq i < j \leq n$.

a) These elements generate W as an R-module. As a matter of fact, since $X_i X_j$ and $Y_i Y_j$ are in I, all polynomials in the $2n$ indeterminates X_i and Y_i can be written modulo I as

$$a_0 + \sum_i a_i X_i + \sum_i b_i Y_i + \sum_{i,j} c_{ij} X_i Y_j.$$

Since $X_i Y_j + X_j Y_i \in I$, we have on one hand $X_i Y_i \in I$ and on the other hand that $X_j Y_i$ are equivalent to $-X_i Y_j$ modulo I. Thus the previous polynomial is equivalent to

$$a_0 + \sum_i a_i X_i + \sum_i b_i Y_i + \sum_{i<j} (c_{ij} - c_{ji}) X_i Y_j$$

or to

$$a_0 + \sum_i a_i X_i + \sum_i b_i Y_i + \sum_{i<j} \frac{1}{2}(c_{ij} - c_{ji})(X_i Y_j - X_j Y_i).$$

Thus its equivalence class (modulo I) is actually a linear combination of the above elements.

b) Let us show that these elements are linearly independent. Suppose that a polynomial

$$a_0 + \sum_i a_i X_i + \sum_i b_i Y_i + \sum_{i<j} c_{ij}(X_i Y_j - X_j Y_i)$$

is in I and is written as

$$\sum_{r,s} \alpha_{rs} X_r X_s + \sum_{r,s} \beta_{rs} Y_r Y_s + \sum_{r \leq s} \gamma_{rs}(X_r Y_s + X_s Y_r)$$

where *a priori* the α, β and γ are polynomials.

- By making all the X and Y equal to zero, we have that $a_0 = 0$.
- By making all the X and Y except one equal to zero, for instance X_i, the result is

$$a_i X_i = \alpha_{ii} X_i^2.$$

For all d in D, we thus get $a_i d = 0$ and thus, with the axiom K-L, $a_i = 0$. In the same way all the b are zero and the polynomial is reduced to

$$\sum_{i<j} c_{ij}(X_i Y_j - X_j Y_i).$$

- Fix an index i and make all the X and Y equal to zero, except X_i and Y_i. Then

$$0 = \alpha_{ii} X_i^2 + \beta_{ii} Y_i^2 + 2\gamma_{ii} X_i Y_i.$$

Let $X_i = d_1$ and $Y_i = d_2$ where d_1 and d_2 are in D. We obtain

$$0 = 2\gamma_{ii}^0 d_1 d_2$$

where γ_{ii}^0 is the independent term of the polynomial γ_{ii}. Thus, by the axiom K-L, $\gamma_{ii}^0 = 0$.

- Finally let us prove that c_{ij} ($i < j$) is null. Let d_1 and d_2 be in D. Assume

$$X_i = d_1, \ Y_i = d_2, \ X_j = -d_1 \text{ and } Y_j = d_2.$$

The identity

$$
\begin{aligned}
c_{ij}(X_iY_j - X_jY_i) &= \alpha_{ii}X_i^2 + \alpha_{ij}X_iX_j + \alpha_{jj}X_j^2 \\
&+ \beta_{ii}Y_i^2 + \beta_{ij}Y_iY_j + \beta_{jj}Y_j^2 \\
&+ 2\gamma_{ii}X_iY_i + 2\gamma_{jj}X_jY_j \\
&+ \gamma_{ij}(X_iY_j + X_jY_i)
\end{aligned}
$$

gives $2c_{ij}d_1d_2 = 0$ because $d_1^2 = d_2^2 = 0$ and $\gamma_{ii}^0 = \gamma_{jj}^0 = 0$. Thus, from the K-L axiom, $c_{ij} = 0$.

We have explicitly indicated a basis for the R-module W ; hence, we see W is a Weil-algebra.

6) It has been verified (Example (3)) that the sum, or tensor product, of two Weil-algebras, is a Weil-algebra. Hence the product of two small objects is a small object. We could also check that a finite limit of small objects is a small object (this notion of finite limit is recalled in 2.2).

2.1.3   The general Kock axiom

Let W be a Weil-algebra. Suppose that W is of breadth r and consider a presentation of W :

$$
W \simeq R[X_1, \ldots, X_r]/I
$$

where I is generated by the polynomials $P_1, P_2, \ldots, P_s$. Let $D(W)$ be the spectrum of W in R, i.e.

$$
D(W) = \{(d_1, d_2, \ldots, d_r) \mid P_i(d_1, \ldots, d_r) = 0, \ 1 \leq i \leq s\}.
$$

We have a homomorphism of R-algebras

$$
R[X_1, \ldots, X_r] \xrightarrow{\alpha} R^{D(W)}
$$

defined by evaluation :

$$
\alpha(P)(d_1, \ldots, d_r) = P(d_1, \ldots, d_r).
$$

This homomorphism passes to the quotient by I because if $Q \in I$ and

$$
(d_1, \ldots, d_r) \in D(W),
$$

then $Q(d_1, \ldots, d_r)$ is null by definition of $D(W)$. We have thus a homomorphism of R-algebras, also denoted by α, or α_W,

$$W \xrightarrow{\alpha} R^{D(W)}.$$

One can easily verify the naturality of α in W, i.e. that the following diagram commutes :

$$
\begin{array}{ccc}
W_1 & \xrightarrow{\alpha_{W_1}} & R^{D(W_1)} \\
\downarrow{\scriptstyle f} & & \downarrow{\scriptstyle R^{D(f)}} \\
W_2 & \xrightarrow[\alpha_{W_2}]{} & R^{D(W_2)}
\end{array}
$$

The general form of the Kock's axiom is as follows :

The general Kock axiom (K-W). *For all Weil-algebras W, the homomorphism*

$$W \xrightarrow{\alpha} R^{D(W)}$$

is bijective.

Examples.

1) The axiom (K-W) applied to

$$W_k = R[X]/(X^{k+1})$$

gives that any mapping f from D_k into R can be written in a unique way as

$$f(\delta) = a_0 + a_1\delta + \ldots + a_k\delta^k.$$

2) In the same way, every mapping from $D(n)$ into R can be written in a unique way as a first degree function.

3) A function from $D \times D$ into R can be written in a unique way as

$$f(d_1, d_2) = a_0 + a_1d_1 + a_2d_2 + a_3d_1d_2.$$

2.2   The general microlinearity of R

2.2.1   The wonderful myopia of R

The objects with shape $D(W)$ where W is a Weil-algebra are so small that R does not see clearly these small objects. As an introduction to this section, we shall give a few examples of this short-sightedness (or, perhaps, far-sightedness) of R.

Let us consider the multiplication

$$D \times D \to D : (d_1, d_2) \rightsquigarrow d_1 \cdot d_2.$$

There is no reason to believe that this multiplication should be surjective. Still, according to a picturesque formulation of Kock :

Proposition 1. *"R thinks that the multiplication : $D \times D \to D$ is surjective". More precisely, if two functions $f_1 : D \to R$ and $f_2 : D \to R$ are equalized by the multiplication "$\cdot$", they are equal. In other words if $f_1(d_1 \cdot d_2) = f_2(d_1 \cdot d_2)$ for all d_1 and d_2 in D, then $f_1(d) = f_2(d)$ for all d in D. Alternatively, if $f(d_1 \cdot d_2) = 0$ for all d_1 and d_2 in D, then f is zero on D.*

Proof: We have $f(d) = a_0 + a_1 d$ and thus $f(d_1 \cdot d_2) = f(0) + a_1 d_1 d_2$. Suppose that we have $f(d_1 \cdot d_2) = 0$. As $f(0) = f(0 \cdot 0) = 0$, we have $a_1 \cdot d_1 \cdot d_2 = 0$ for all d_1 and d_2 in D hence $a_1 = 0$. So f is identically zero.   ∎

In the same way the addition

$$D \times D \xrightarrow{+} D_2 : (d_1, d_2) \rightsquigarrow d_1 + d_2$$

does not have any reason to be surjective. Still

Proposition 2. *"R thinks that the addition $+ : D \times D \to D_2$ is surjective". More precisely if two functions $f_1 : D_2 \to R$ and $f_2 : D_2 \to R$ are equalized by "$+$", they are equal. In other words, if $f : D_2 \to R$ is such that $f(d_1 + d_2) = 0$ for all d_1 and d_2 in D, then $f(\delta) = 0$ for all δ in D_2.*

Proof: Let $f : D_2 \to R$. It is written in a unique way as

$$f(\delta) = a_0 + a_1\delta + a_2\delta^2.$$

Let $g : D \times D \to R : (d_1, d_2) \rightsquigarrow g(d_1, d_2) = f(d_1 + d_2)$. We have, in a unique way,

$$g(d_1, d_2) = \alpha_0 + \alpha_1 d_1 + \alpha_2 d_2 + \alpha_3 d_1 \cdot d_2.$$

So we have the identity

$$a_0 + a_1(d_1 + d_2) + a_2(d_1 + d_2)^2 = \alpha_0 + \alpha_1 d_1 + \alpha_2 d_2 + \alpha_3 d_1 \cdot d_2,$$

from which it follows

- by putting $d_1 = d_2 = 0$, that $a_0 = \alpha_0$,

- and putting $d_2 = 0$, we have $a_1 d_1 = \alpha_1 d_1$ for all d_1 in D and thus $a_1 = \alpha_1$; by putting $d_1 = 0$, we have $a_1 d_2 = \alpha_2 d_2$ for all d_2 in D and so $a_1 = \alpha_2$,

- thus $2a_2 d_1 d_2 = \alpha_3 d_1 d_2$ for all d_1 and d_2 in D and so $\alpha_3 = 2a_2$.

If g is zero, we have $\alpha_0 = \alpha_1 = \alpha_2 = \alpha_3 = 0$ and so $a_0 = a_1 = a_2 = 0$ and f is identically zero. ∎

In order to state other similar facts, let us fix some terminology. Consider the diagram

$$A \underset{v}{\overset{u}{\rightrightarrows}} B \overset{w}{\longrightarrow} C \qquad\qquad (1)$$

with $w \circ u = w \circ v$ (in a category C). We say that an object X perceives (1) as a co-equalizer if every arrow $f : B \to X$ such that $f \circ u = f \circ v$ factorizes in a unique way by W, i.e. there exists one and only one arrow $g : C \to X$ such that $g \circ w = f$. Note that (1) is a co-equalizer if all objects of C perceive (1) as a coequalizer. If we denote by X^A the "set" (of our naive intuitionistic set theory, and we suppose that this "set" X^A exists) of arrows from A to X, to say that X perceives (1) as a coequalizer amounts to saying that

$$X^C \xrightarrow{\ X^w\ } X^B \underset{X^v}{\overset{X^u}{\rightrightarrows}} X^A \qquad (2)$$

(2) is an equalizer in our category of sets, i.e. X^C corresponds bijectively, by X^w, to the set of elements of X^B having the same image by X^u and X^v.

Consider in the same way a commutative square :

$$
\begin{array}{ccc}
A & \xrightarrow{\ u\ } & B \\
{\scriptstyle v}\downarrow & & \downarrow{\scriptstyle v'} \\
C & \xrightarrow[\ u'\ ]{} & D
\end{array}
\qquad (3)
$$

$(u' \circ v = v' \circ u)$. We say that an object X perceives (3) as a *pushout* if for any arrow $x : C \to X$ and $y : B \to X$ such that $x \circ v = y \circ u$, there exists one and only one arrow $z : D \to X$ such that $z \circ u' = x$ and $z \circ v' = y$. So (3) is a pushout if every object of C perceives (3) as a pushout. To say that X perceives (3) as a pushout is to say that

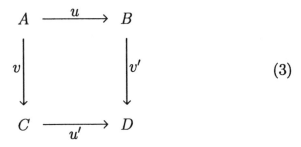

$$
\begin{array}{ccc}
X^A & \xleftarrow{\ X^u\ } & X^B \\
{\scriptstyle X^v}\uparrow & & \uparrow{\scriptstyle X^{v'}} \\
X^C & \xleftarrow[\ X^{u'}\ ]{} & X^D
\end{array}
$$

is a pullback (of "sets"), i.e. X^D is in bijection with the set of pairs of elements of X^B and X^C having the same image in X^A.

Proposition 3. *Let us consider the diagram*

$$D \times D \underset{\tau}{\overset{id}{\rightrightarrows}} D \times D \xrightarrow{\ +\ } D_2 \qquad (1)$$

where id is the identity and τ is the twist $\tau(d_1, d_2) = (d_2, d_1)$. I claim that R perceives this diagram as a coequalizer.

Proof: First (1) commutes because $d_1 + d_2 = d_2 + d_1$. Let

$$f : D \times D \to R.$$

This function is written in a unique way as

$$f(d_1, d_2) = a_0 + a_1 d_1 + a_2 d_2 + b d_1 d_2.$$

If f equalizes id and τ, in other words if f is symmetrical, $f(d_1, d_2) = f(d_2, d_1)$, and we have $f(d, 0) = f(0, d)$ for all d in D. So $a_1 d = a_2 d$ for all d in D, hence $a_1 = a_2$. Let us define then

$$g : D_2 \to R$$

by setting

$$g(\delta) = f(0, 0) + a_1 \delta + \frac{b}{2} \delta^2.$$

We have

$$
\begin{aligned}
g(d_1 + d_2) &= f(0, 0) + a_1 \cdot (d_1 + d_2) + \frac{b}{2}(d_1 + d_2)^2 \\
&= f(d_1, d_2).
\end{aligned}
$$

On the other hand, since R thinks that "+" is surjective, g is uniquely determined. ∎

The proof shows also that R perceives $D \xrightarrow{i_1} D \times D \xrightarrow{+} D_2$ as a coequalizer (where $i_1(d) = (d, 0)$ and $i_2(d) = (0, d)$).

Proposition 4. *Let us consider the diagram*

$$D \underset{0}{\overset{i}{\rightrightarrows}} D_2 \xrightarrow{q} D \qquad\qquad (1)$$

where i is the canonical injection, 0 the constant function zero and $q(\delta) = \delta^2$. I claim that R perceives (1) as a coequalizer.

Proof: First (1) commutes because the elements of D are of zero square. So let $f : D_2 \to R$. We have, in a unique way,

$$f(\delta) = f(0) + a_1 \delta + a_2 \delta^2.$$

So we have, if $d \in D$,

$$f(d) = f(0) + a_1 d$$

and if $f \circ i = 0$, we have $a_1 d = 0$ for all d in D and thus $a_1 = 0$.

Let us define then $g : D \to R$ by setting

$$g(d) = f(0) + a_2 d;$$

we have

$$g(q(\delta)) = g(\delta^2) = f(0) + a_2 \delta^2 = f(\delta).$$

There remains the unicity of g to be verified. We shall show then that R thinks that q is surjective. Suppose that $g : D \to R$ is such that $g \circ q = 0$. We have $g(d) = g(0) + a_1 d$ with $g(\delta^2) = 0$. First $g(0) = 0$. Then let d_1 and d_2 be in D. We have

$$
\begin{aligned}
0 = g((d_1 + d_2)^2) &= a_1 \cdot (d_1 + d_2)^2 \\
&= 2a_1 d_1 d_2
\end{aligned}
$$

and so $a_1 = 0$. Hence g is null. ∎

Proposition 5. *Let us consider the diagram*

$$D \times D \times D \underset{id \times m}{\overset{m \times id}{\rightrightarrows}} D \times D \xrightarrow{\quad m \quad} D \qquad (1)$$

where m is the multiplication. I say that R perceives (1) as a coequalizer.

Proof: The commutativity of the diagram is simply the associativity of the multiplication m. Let $f : D \times D \to R$. We have

$$f(d, d') = a_0 + a_1 d + a_2 d' + a_3 dd' \qquad (2)$$

If f equalizes $m \times id$ and $id \times m$, we have, for all d_1, d_2 and d_3 in D, that $f(d_1 d_2, d_3) = f(d_1, d_2 d_3)$. By setting $d_1 = d_2 = 0$ in this identity and by using (2), this becomes

$$a_0 + a_2 d_3 = a_0$$

for all d_3 in D, and thus $a_2 = 0$. In the same way, by doing $d_2 = d_3 = 0$ we find $a_1 = 0$. If we put $g : D \to R : d \rightsquigarrow a_0 + a_3 d$, we get $g(d \cdot d') = f(d, d')$. The uniqueness of g follows from the fact that R thinks that the multiplication is surjective (Proposition 1). ∎

Consider the diagram

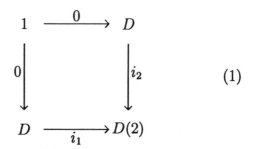

$$(1)$$

where 1 is the singleton, 0 denotes the null mapping and i_1 and i_2 are canonical injections : $i_1(d) = (d, 0)$, $i_2(d) = (0, d)$. It is obviously a commutative square. Let us consider the diagram

$$(2)$$

where 0 denotes null functions, i_1 is the injection given by

$$i_1(d_1, d_2, \ldots, d_p) = (d_1, d_2, \ldots, d_p, 0, 0, \ldots, 0)$$

(which actually is in $D(p+q)$) and i_2 is the injection

$$i_2(d_1, d_2, \ldots, d_q) = (0, 0, \ldots, 0, d_1, d_2, \ldots, d_q).$$

Proposition 6. *R perceives the diagram (1) and the diagrams (2) as pushouts.*

Proof: Let

$$f : D(p) \to R \quad \text{and} \quad g : D(q) \to R$$

with $f(0_p) = g(0_q)$ which we denote a_0. These functions are uniquely expressed as first degree function (cf. 1.3, Example 2), hence we write

$$\begin{aligned} f(d_1, d_2, \ldots, d_p) &= a_0 + a_1 d_1 + a_2 d_2 + \ldots + a_p d_p \\ g(d'_1, d'_2, \ldots, d'_q) &= a_0 + b_1 d'_1 + b_2 d'_2 + \ldots + b_q d'_q. \end{aligned}$$

We define $h : D(p+q) \to R$ by

$$h(d_1, d_2, \ldots, d_p \ , \ d_{p+1}, \ldots, d_{p+q})$$
$$= a_0 + a_1 d_1 + \ldots + a_p d_p + b_1 d_{p+1} + \ldots + b_q d_{p+q'}.$$

We have $h(i_1(\underline{d})) = f(\underline{d})$ and $h(i_2(\underline{d'})) = g(\underline{d'})$. Let us check the unicity. Let $k : D(p+q) \to R$ with $k \circ i_1 = f$ and $k \circ i_2 = g$. We have

$$k(d_1, \ldots, d_{p+q}) = \alpha_0 + \alpha_1 d_1 + \ldots + \alpha_{p+q} d_{p+q}.$$

By putting every d to be zero, we get $\alpha_0 = a_0$. By putting every d but one to be zero, we find equality of coefficients of d_i in k and in h and thus the desired unicity. ∎

Proposition 7. *R perceives the diagram*

$$D \xrightarrow[\substack{0 \\[-2pt] \longrightarrow}]{\substack{i_1 \\[-2pt] \longrightarrow \\[-2pt] i_2}} D \times D \xrightarrow{\ m\ } D$$

(where $i_1(d) = (d, 0)$, $i_2(d) = (0, d)$, $0(d) = (0, 0)$, $m(d_1, d_2) = d_1 \cdot d_2$) as a (double) coequalizer.

Proof: Let $f : D \times D \to R$ with $f(d, 0) = f(0, 0) = f(0, d)$ (for every d in D). We have

$$f(d_1, d_2) = f(0, 0) + a_1 d_1 + a_2 d_2 + a_3 d_1 d_2.$$

By making d_1 (resp. d_2) zero, we find that a_2 (resp. a_1) is null. We define $g : D \to R$ by $g(d) = f(0, 0) + a_3 d$ and g is in fact the only mapping such that $g(d_1 \cdot d_2) = f(d_1, d_2)$. ∎

One can of course indefinitely multiply this kind of example. As an exercise, the reader could verify for instance that R perceives :

$$D_2 \times D_2 \times D_2 \xrightarrow[\substack{id \times m}]{\substack{m \times id \\ \longrightarrow \\ \longrightarrow}} D_2 \times D_2 \xrightarrow{\ m\ } D_2$$

as a coequalizer, where m is the multiplication.

- Let $(D \times D) \vee D = \{(d_1, d_2, d_3) \mid d_i \in D, \ d_1 d_3 = d_2 d_3 = 0\}$. R perceives

$$I \xrightarrow{\quad 0 \quad} D$$

$$\Big\downarrow 0 \qquad\qquad \Big\downarrow \epsilon$$

$$D \times D \xrightarrow{\quad \varphi \quad} (D \times D) \vee D$$

(where $(d) = (0,0,d)$ and $\varphi(d_1,d_2) = (d_1,d_2,0)$) as a pushout.

- R perceives

$$D_k \times D_k \times D_k(n) \mathrel{\substack{m \times id \\ \xrightarrow{\hspace{1.3cm}} \\ \xrightarrow{\hspace{1.3cm}} \\ id \times m'}} D_k \times D_k(n) \xrightarrow{\quad m' \quad} D_k(n)$$

as a coequalizer. Here we put

$$D_k(n) = \{(d_1,\ldots,d_n) \,|\, d_{i_1} \cdot d_{i_2} \cdots d_{i_k} = 0 \; \forall i_1, i_2, \ldots, i_k\}$$

$m : D_k \times D_k \to D_k$ is the product, $m' : D_k \times D_k(n) \to D_k(n)$ is given by $m'(\delta, d_1, \ldots, d_n) = (\delta d_1, \delta d_2, \ldots, \delta d_n)$.

2.2.2   Quasi colimits of small objects

The different propositions proved in the previous section show sufficient similarity to suggest that there must be a general underlying phenomenon.

Let $R - Alg$ be the category whose objects are the R-algebras and whose arrows are homomorphisms of R-algebras.

We recall the meaning of *finite limit* in $R - Alg$.

Let $\mathcal{D}$ be a finite diagram of $R - Alg$ (i.e. a finite family of objects $A_1, A_2, \ldots, A_k$ and between them a finite set of arrows). Consider the set $L_D \subseteq A_1 \times A_2 \times \ldots \times A_k$,

$$L_D = \{(a_i)_{1 \leq i \leq k} \mid \text{ for every } f : A_i \to A_j \text{ in } D, \; f(a_i) = a_j\}$$

As each $f : A_i \to A_j$ in $\mathcal{D}$ is a homomorphism, it is easy to see that L_D is a sub R-algebra of the product and the projections $L_D \xrightarrow{\lambda_i} A_i$ are homomorphisms. We say that L_D is *the limit of the diagram $\mathcal{D}$*.

One verifies that $(L_\mathcal{D} \xrightarrow{\lambda_i} A_i)_{1 \le i \le k}$ is characterized, up to isomorphism, by the following universal property :

(a) for every $f : A_i \to A_j$ in $\mathcal{D}$, the triangle

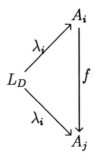

commutes (we then say that $(L_\mathcal{D} \xrightarrow{\lambda_i} A_i)$ is a left-cone of vertex $L_\mathcal{D}$ and of base $\mathcal{D}$).

(b) Every left-cone of vertex M and of base $\mathcal{D}$, $(M \xrightarrow{\mu_i} A_i)_{1 \le i \le k}$, factorizes in a unique way by $L_\mathcal{D}$. In other words, there exists one and only one homomorphism $\mu : M \to L$ such that, for every i, $\lambda_i \circ \mu = \mu_i$.

We are interested in the case where the A_i of $\mathcal{D}$ and $L_\mathcal{D}$ are Weil-algebras.

Notice that even if the A_i are Weil-algebras, this does not ensure that $L_\mathcal{D}$ be a Weil-algebra. Let us consider for instance the diagram simply made up of two Weil-algebras

$$A_1$$

$$A_2$$

The limit (on the left) of this diagram in $R - Alg$ is simply the set product $A_1 \times A_2$ with its product stucture of R-algebras and projections

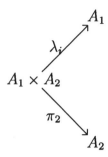

But the set product of two Weil-algebras is not a Weil-algebra. If, as an R-module, we have $A_1 = R^{n+1}$ and $A_2 = R^{p+1}$, the product in the category of R-algebras is $A_1 \times A_2 \simeq R^{n+p+2}$. On the other hand, the Weil-algebra "product" in the category of Weil-algebras is, as an R-module, R^{n+p+1}.

Let $\mathcal{D}$ be a finite diagram in the category of Weil-algebras. Let $(L \xrightarrow{\lambda_i} A_i)$ be its limit in the category of R-algebras. We say that it is a *good finite limit* (on the left) of Weil-algebras if L is a Weil-algebra and λ_i are homomorphisms of Weil-algebras. It is easy to see that a good finite limit of Weil-algebras is *a fortiori* a limit in the category of Weil-algebras.

Let us then apply to these good limits the contravariant functor $D = Spec_R$ from the category of Weil-algebras to the one of sets.

Definition 1. *Let* $(L \xrightarrow{\lambda_i} A_i)$ *be a good finite limit of Weil-algebras. The diagram obtained by applying the functor* D

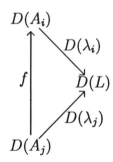

is called a finite quasi colimit *of small objects.*

As an example, we shall indicate now the way to get the diagrams of the Propositions 3, 4, 5, 6 and 7 as quasi colimits of small objects. 1.-Let us consider the diagram

$$D \times D \underset{\tau}{\overset{id}{\rightrightarrows}} D \times D \xrightarrow{\ +\ } D_2 \qquad (1)$$

(cfr. Proposition 3). We have the diagram

$$R[X] \xrightarrow{\ +'\ } R[X,Y] \underset{\tau'}{\overset{id'}{\rightrightarrows}} R[X,Y] \qquad (1')$$

where $+'(X) = X + Y$; $id' = id$; $\tau'(X) = Y$ and $\tau'(Y) = X$. Passing to the quotient we get the diagram (1") :

$$R[X]/(X^2) \to R[X,Y]/(X^2,Y^2) \rightrightarrows R[X,Y]/(X^2,Y^2)$$

where the first arrow is denoted by $+''$ and the parallel arrows by id'' and τ''

One verifies then easily that

(a) (1") is an equalizer in the category of R-algebras;

(b) (1") is in the category of Weil-algebras;

(c) the functor $D(\cdot)$ transforms (1") into (1).

2.-Let us consider the diagram

$$D \underset{0}{\overset{i}{\rightrightarrows}} D_2 \xrightarrow{\ q\ } D \qquad (2)$$

(cfr. Proposition 4) where $q(\delta) = \delta^2$. We consider

$$R[X] \xrightarrow{\ q'\ } R[X] \underset{0'}{\overset{i'}{\rightrightarrows}} R[X] \qquad (2')$$

with $q'(X) = X^2$, $i'(X) = X$, $D'(X) = 0$ and we pass to the quotient :

$$R[X]/(X^2) \xrightarrow{q''} R[X]/(X^3) \underset{0''}{\overset{i''}{\rightrightarrows}} R[X]/(X^2) \qquad (2'')$$

One verifies also that (2") is an equalizer in the category of R-algebras, that (2") is in the category of Weil-algebras and that the functor $D(\cdot)$ transforms (2") into (2).

3.-Let us consider the diagram

$$D \times D \times D \underset{id \times m}{\overset{m \times id}{\Longrightarrow}} D \times D \xrightarrow{\quad m \quad} D \tag{3}$$

(cf. Proposition 5) where m is the multiplication. We have the diagram

$$R[X] \xrightarrow{\quad m' \quad} R[X,Y] \underset{m'_{23}}{\overset{m'_{12}}{\Longrightarrow}} R[X,Y,Z] \tag{3'}$$

given by $m'(X) = X \cdot Y$; $m'_{12}(X) = X \cdot Y$ and $m'_{12}(Y) = Z$; $m'_{23}(X) = X$ and $m'_{23}(Y) = Y \cdot Z$. By passing to the quotient we construct the diagram (3") :

$$R[X]/(X^2) \to R[X,Y]/(X^2,Y^2) \rightrightarrows R[X,Y,Z]/(X^2,Y^2,Z^2)$$

and draw conclusions as above.

4.-Let us consider the diagram

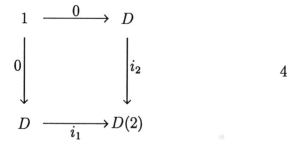

(cf. Proposition 6). It follows from the indicated technique (passing to the quotient and applying $D(\cdot)$) applied to the diagram

$$
\begin{array}{ccc}
R & \xleftarrow{\quad 0' \quad} & R[X] \\
{\scriptstyle 0'}\big\uparrow & & \big\uparrow{\scriptstyle i''_2} \\
R[X] & \xleftarrow[\;\;i''_1\;\;]{} & R[X,Y]
\end{array}
\tag{4'}
$$

where $0'(X) = 0$; $i''_1(X) = X$ and $i''_1(Y) = 0$; $i''_2(X) = 0$ and $i''_2(Y) = X$. In the same way

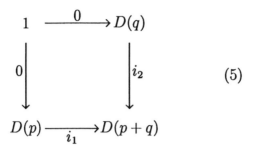

$$1 \xrightarrow{\quad 0 \quad} D(q)$$

with vertical maps 0 and i_2, to $D(p) \xrightarrow{\quad i_1 \quad} D(p+q)$. (5)

follows from

$$R \xleftarrow{\quad 0' \quad} R(Y_1, Y_2, \ldots, Y_q)$$

with vertical maps $0'$ and i_2', (5')

$$R(X_1, X_2, \ldots, X_p) \xleftarrow{\quad i_1' \quad} R(X_1, \ldots, X_p, Y_1, \ldots, Y_q)$$

5.-The diagram

$$D \overset{\overset{i_1}{\underset{i_2}{\xrightarrow{\hspace{1cm}}}}}{\underset{0}{\xrightarrow{\hspace{1cm}}}} D \times D \xrightarrow{\quad m \quad} D$$

(cfr. Proposition 7) follows in the same way from

$$R[X] \xrightarrow{\quad m' \quad} R[X, Y] \overset{\overset{i_1'}{\underset{i_2'}{\xrightarrow{\hspace{1cm}}}}}{\underset{0'}{\xrightarrow{\hspace{1cm}}}} R[X]$$

where $m'(X) = X \cdot Y$; $i_1'(X) = X$ and $i_1'(Y) = 0$; $i_2'(X) = 0$ and $i_2'(Y) = X$; $0'(X) = 0'(Y) = 0$.

2.2.3 Perception by R of quasi-colimits of small objects

The various examples of quasi-colimits that have been quoted show that the propositions of section 2.1 are implied by the following proposition.

Proposition 8. *If R verifies the general Kock axiom (K-W), then R perceives the finite quasi-colimits of small objects as colimits.*

Proof: If R satisfies the axiom (K-W), W is isomorphic, as an R-algebra, to $R^{D(W)}$. If we have a good limit of Weil-algebras $(L \xrightarrow{\lambda_i} W_i)$, it is a limit in the category of R-algebras and thus

$$R^{D(L)} \xrightarrow{R^{D(\lambda_i)}} R^{D(W_i)}$$

is a limit of R-algebras. But this expresses the fact that R perceives $(D(W_i) \xrightarrow{D(\lambda_i)} D(L))$ as a colimit. ∎

That is precisely the content of the short-sightedness of R that we observed in 2.1. This shows that the propositions in this section were superfluous in that they are contained in Proposition 8.

We shall see now on the contrary that these propositions in 2.1 and other similar ones are very useful. Actually it is not always quite straightforward to show that a cone (on the left) of Weil algebras is a limit (and a good limit). Thus, it is not quite straightforward to show that a diagram of small objects (in the shape of a right-cone) is a quasi-colimit. The following criterion is often more practical than the definition. We suppose that R satisfies (K-W).

Proposition 9. *Given a finite cone (on the right) of small objects, i.e. the image by $D(\cdot)$ of a (finite) cone (on the left)*

$$(L \xrightarrow{\lambda_i} W_i)$$

in the category of Weil-algebras, if R perceives it as a colimit, it is a quasi colimit.

Proof: If R perceives the diagram

$$((D(W_i) \xrightarrow{D(\lambda_i)} D(L))$$

as a colimit, then

$$R^{D(L)} \xrightarrow{R^{D(\lambda_i)}} R^{D(W_i)}$$

actually is a limit in "Set" and thus a limit in the category of R-algebras. Now, according to the axiom (K-W) this is the diagram

$$(L \xrightarrow{\lambda_i} W_i)$$

which is then a good finite limit of Weil algebras, and the proposition is proved. ∎

2.3 General microlinearity

2.3.1 The notion of general microlinearity

We suppose from now on that R satisfies the axiom (K-W).

Definition 1. *An object M is said to be* microlinear *if it perceives the finite quasi-colimits of small objects as colimits.*

Proposition 8 of 2.2 can be stated as : R is microlionear. Actually, by taking R, we can construct many microlinear objects.

Proposition 1. *Any limit of microlinear objects is microlinear. If M is microlinear and X is any "set", M^X is microlinear.*

Proof: Let $(M \xrightarrow{\mu_\alpha} M_\alpha)_\alpha$ be a limit where the M_α are microlinear. Let $(D(W_i) \xrightarrow{\lambda_i} D(L))$ be a finite quasi-colimit of small objects. Since every M_α is microlinear,

$$(M_\alpha^{D(L)} \xrightarrow{M^{\lambda_i}} M^{D(W_i)})_i$$

is a limit. Thus M is microlinear. This proof is also valid for M^X. ∎

For instance R^n is microlinear but also R^X. An equalizer on the left of $f, g : R^n \to R$, for example

$$\{x \in R^n \mid f(x) = 0\}$$

is microlinear ; thus all usual manifolds are microlinear. But we have also objects which would usually be considered as being of infinite dimension (for example $M_1^{M_2}$: the object of functions of M_2 into M_1). Or objects that have singularities like for instance

$$\{(x,y) \in R^2 \mid y^2 - x^3 - x^2 = 0\}$$

(but let us not forget that this "curve" contains for instance $D \times D$).

2.3.2 A few elementary consequences of general microlinearity

We intend to show here that a certain number of conditions or axioms that have been postulated in the literature, needed for the development of synthetic differential geometry, are included in the property of general microlinearity.

1) An object M is *k-infinitesimally linear* $(k \in \mathbf{N})$ if for all functions $t_1, t_2, \ldots, t_k : D \to M$ such that $t_1(0) = \ldots = t_k(0)$, there exists one and only one function $\ell : D(k) \to M$ whose restrictions to the axes coincide with the t_r, i.e. $\ell \circ i_r = t_r$ $(r = 1, 2, \ldots, k)$ where i_r is the r-th injection $D \to D(k) : d \rightsquigarrow (0, 0, \ldots, d, 0, \ldots, 0)$.

Proposition 2. *If M is microlinear, it is k-infinitesimally linear.*

Proof: This follows immediately from the fact that the diagrams of Proposition 6 in 2.2 are quasi-colimits.   ∎

2) An object M verifies the *Wraith axiom W* if for all function $\tau : D \times D \to M$ constant on the axes (i.e. $\tau(d, 0) = \tau(0, d) = \tau(0, 0)$ for every d in D) there exists one and only one function $t : D \to M$ such that for all $(d_1, d_2) \in D \times D$, $t(d_1 \cdot d_2) = \tau(d_1, d_2)$.

Proposition 3. *If M is microlinear, it satisfies the axiom W.*

Proof: It suffices to observe that the diagram of Proposition 7, 2.2, is a quasi-colimit.   ∎

3) An object M satisfies the *Bunge axiom B* (or axiom of iterated tangent bundle) if for every function $\tau : D \times D \to M$ such that for each d in D, $f(0, d) = f(d, 0)$, there exists one and only one function $g : D_2 \to M$ such that for every $(d_1, d_2) \in D \times D$, $\tau(d_1, d_2) = g(d_1 + d_2)$.

Proposition 4. *If M is microlinear, it satisfies axiom B.*

Proof: The diagram

$$ D \overset{i_1}{\underset{i_2}{\rightrightarrows}} D \times D \overset{+}{\longrightarrow} D_2 $$

is a quasi-colimit (cfr. Proposition 3, section 2.2).   ∎

We shall use other particular cases of microlinearity in what follows.

2.4 Commented bibliography

1) The concept of Weil-algebra was introduced by Weil in [56] (1953). It was extensively used by Dubuc [14] (1979) for the construction of models of synthetic differential geometry. One can find other developments on Weil-algebra by Dubuc and Reyes in [16] (1979). The axiom (K-W) was fomulated by Kock in [31] (1981).

2)The general notion of infinitesimally linear objects was introduced by F. Bergeron [4] (1980). At the level of the theory itself, they are what, in this book, are called microlinear objects. This notion was further analysed and used by Kock and Lavendhomme in [35] (1984).

3) The idea of infinitesimal linearity (cf. 2.3.2) which is necessary for the study of tangent spaces, emerged in discussions between Kock, Reyes and Wraith, as recounted by Kock in his book [31] ; it was then used by Reyes and Wraith [54] (1978). The Wraith's axiom W also comes from [54] and is used by Kock in [30] (1979).

The classical study of iterated tangent bundles is familiar to geometers. One may read, for instance, Claude Godbillon [21] (1969). In synthetic differential geometry, the axiom of iterated tangent bundles is explored by Bunge and P. Sawyer in [10] (1984).

Chapter 3

Tangency

3.1 Tangent bundle

3.1.1 The tangent module at a point

Let M be a microlinear object and m a point in M. Classically, a tangent vector at a point m in a differentiable manifold M can be described as an equivalence class of curves passing through m, two arcs being considered equivalent if their expressions in local coordinates have the same derivatives. But here, as we look at things in the infinitesimal - we are in the infinitesimal context - we can identify a tangent vector with an infinitesimal shifting, in other words with a "micro-arc" defined only on D.

Definition 1. *A vector tangent to M at m is a mapping*

$$t : D \to M$$

such that $t(0) = m$.

The set of tangent vectors to M at m comes equipped with an R-module structure. The product of a tangent vector t by a scalar α is defined by $(\alpha \cdot t)(d) = t(\alpha \cdot d)$ for every d of D. Let us define now the sum of two tangent vectors at m, t_1 and t_2. We know that R perceives the diagram

$$1 \xrightarrow{\quad 0 \quad} D$$
$$0 \downarrow \qquad \downarrow i_2$$
$$D \xrightarrow[\quad i_1 \quad]{} D(2)$$

as a pushout (2.2, Prop. 6). This is also true for M as M is microlinear. As $t_1(0) = t_2(0) = m$, there is one and only one mapping from $D(2)$ into M, denoted $\ell_{(t_1,t_2)}$, such that $\ell_{(t_1,t_2)}(d,0) = t_1(d)$ and $\ell_{(t_1,t_2)}(0,d) = t_2(d)$ for every d in D. So we define

$$t_1 + t_2 : D \to M$$

with $(t_1 + t_2)(d) = \ell_{(t_1,t_2)}(d,d)$.

We must define the vector zero, 0, by $0(d) = m$ for every d of D and the opposite $-t$ of a vector t by $(-t)(d) = t(-d)$.

Proposition 1. *With the above mentioned operations the set $T_m M$ of tangent vectors to M at m has a unitary R-module structure.*

Proof:

a) We have $t + 0 = t$. Actually the function $\ell_{(t,0)} : D(2) \to M$ is uniquely characterized by $\ell_{(t,0)}(d,0) = t(d)$ and $\ell_{(t,0)}(0,d) = m$. As the function $t \circ \pi_1 : D(2) \to M : (d_1,d_2) \rightsquigarrow t(d_1)$ satisfies these two identities, we have

$$(t + 0)(d) = \ell_{(t,0)}(d,d) = (t \circ \pi_1)(d,d) = t(d)$$

b) Let us verify the associativity of addition. According to Proposition 6 (2.2), the diagrams

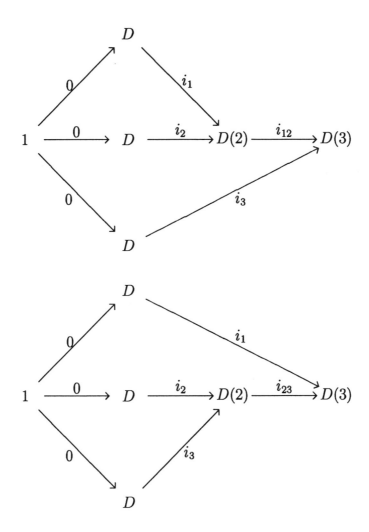

(where $i_{12}(d_1, d_2) = (d_1, d_2, 0)$ and $i_{23}(d_2, d_3) = (0, d_2, d_3)$) are perceived by R as being pushouts. This is also true for M which is microlinear. Let $\ell : D(3) \to M$ be the unique function such that $\ell(d, 0, 0) = t_1(d)$, $\ell(0, d, 0) = t_2(d)$ and $\ell(0, 0, d) = t_3(d)$. So we have $\ell(d_1, d_2, 0) = \ell_{(t_1, t_2)}(d_1, d_2)$ and $\ell(0, d_2, d_3) = \ell_{(t_2, t_3)}(d_2, d_3)$. Hence $((t_1 + t_2) + t_3)(d) = \ell(d, d, d) = (t_1 + (t_2 + t_3))(d)$.

c) We have $t + (-t) = 0$. In fact the function $\ell : D(2) \to M$, $\ell(d_1, d_2) = t(d_1 - d_2)$ is such that $\ell(d, 0) = t(d)$ and $\ell(0, d) = t(-d) = (-t)(d)$ and so $(t + (-t))(d) = \ell(d, d) = t(0) = m$.

d) The equality $\ell_{(t_2,t_1)}(d_2, d_1) = \ell_{(t_1,t_2)}(d_1, d_2)$ being true on the axes, is true everywhere in $D(2)$ and in particular on the diagonal, and thus the addition is commutative.

e) Considering

$$\lambda(d_1, d_2) = t(\alpha d_1 + \beta d_2)$$

gives us

$$(\alpha + \beta)t = \alpha t + \beta t.$$

Considering

$$\lambda(d_1, d_2) = \ell_{(t_1,t_2)}(\alpha d_1, \alpha d_2)$$

gives us

$$\alpha \cdot (t_1 + t_2) = \alpha t_1 + \alpha t_2.$$

The identities $\alpha \cdot (\beta \cdot t) = (\alpha \cdot \beta) \cdot t$ and $1 \cdot t = t$ being immediate, the proposition is proved.

∎

Proposition 2. *The R-module $T_m M$ is Euclidean.*

Proof: Let $\varphi : D \to T_m M$; we must prove (cf. 1.1.4) that there is one and only one vector $t \in T_m M$ such that

$$\varphi(d) = \varphi(0) + d \cdot t$$

for every d in D.

Let us consider the function $\tau : D \times D \to M$ given by

$$\tau(d_1, d_2) = (\varphi(d_1) - \varphi(0))(d_2).$$

We have

$$\tau(d_1, 0) = (\varphi(d_1) - \varphi(0))(0) = m,$$

because $\varphi(d_1) - \varphi(0)$ is in $T_m M$. We have also

$$\tau(0, d_2) = (\varphi(0) - \varphi(0))(d_2) = m.$$

As (2.2, Proposition 7) the diagram

$$D \xrightarrow[\substack{i_1 \\ i_2 \\ 0}]{} D \times D \xrightarrow{m} D$$

is a quasi colimit and M is microlinear (in other words M satisfies axiom (W) : 2.3, Proposition 3), there is one and only one function $t : D \to M$ such that

$$\tau(d_1, d_2) = t(d_1 \cdot d_2).$$

We have then

$$
\begin{aligned}
(\varphi(d_1) - \varphi(0))(d_2) &= t(d_1 \cdot d_2) \\
&= (d_1 \cdot t)(d_2)
\end{aligned}
$$

and so

$$\varphi(d_1) = \varphi(0) + d_1 \cdot t.$$

The uniqueness of the vector t satisfying this property results immediately from the uniqueness of the factorisation of $\tau : D \times D \to M$ through the multiplication $m : D \times D \to D$. ∎

Let $f : M_1 \to M_2$ be a mapping of one microlinear object into another microlinear object. Let $m \in M_1$. We define

$$df_m : T_m M_1 \to T_{f(m)} M_2$$

merely by composition :

$$df_m(t) = f \circ t.$$

Proposition 3. *The mapping*

$$df_m : T_m M_1 \to T_{f(m)} M_2$$

is linear.

Proof: The mapping df_m obviously satisfies :

$$
\begin{aligned}
df_m(\alpha \cdot t)(d) &= f(t(\alpha \cdot d)) \\
&= (\alpha \cdot df_m(t))(d)
\end{aligned}
$$

As the R-module $T_{f(m)} M_2$ is Euclidean (Proposition 2), df_m is linear by 1.2, Proposition 10. ∎

Let us denote by Mic_b the category whose objects are the couples (M, m) of a microlinear object M and a point m of M ; a morphism $f : (M_1, m_1) \to (M_2, m_2)$ is a mapping of M_1 into M_2 such that $f(m_1) = m_2$. Let Euc be the category of Euclidean R-modules and R-linear applications. We have a functor

$$T : Mic_b \to Euc$$

defined by

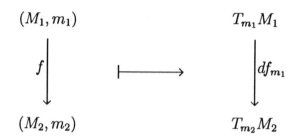

$$
\begin{array}{ccc}
(M_1, m_1) & & T_{m_1}M_1 \\
\downarrow f & \longmapsto & \downarrow df_{m_1} \\
(M_2, m_2) & & T_{m_2}M_2
\end{array}
$$

Functoriality character is immediate. For instance if g and f are composable

$$d(g \circ f)_{m_1}(t) = (g \circ f) \circ t = g \circ (f \circ t) = dg_{m_2}(df_{m_1}(t)).$$

The functor T is called the tangency functor.

3.1.2   Vector bundle and tangent bundle

Let $\xi : E \to M$ be a mapping of microlinear objects. We put, for $m \in M$, $\xi_m = \xi^{-1}(\{m\})$ and we say that ξ_m is the fibre of ξ at m.

Definition 2. *We say that ξ is a vector bundle if for all m in M, ξ_m is an R-module. We say that ξ is a Euclidean vector bundle if each ξ_m is an Euclidean R-module.*

The mapping

$$\tau : M^D \to M$$

sending each tangent vector $t : D \to M$ to its contact point $\tau(t) = t(0)$ is called the tangent bundle on M. The following proposition is immediate.

Proposition 4. *The mapping $\tau : M^D \to M$ is a Euclidean vector bundle.*

Proof: As M is microlinear, M^D is microlinear (Proposition 1) ; then τ is a Euclidean vector bundle according to Propositions 1 and 2 here above. ∎

Consider a commutative square

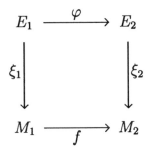

where ξ_1 and ξ_2 are vector bundles. The commutativity of the square tells us that φ takes the fibre of ξ_1 in m into the fibre of ξ_2 in $f(m)$. We denote by φ_m this mapping of $(\xi_1)_m$ into $(\xi_2)_{f(m)}$. We say that (φ, f) is a morphism of ξ_1 to ξ_2 if, for all m of M_1, φ_m is linear.

If $f : M_1 \to M_2$ is a mapping (in the category *Mic* of microlinear objects), we have the commutative diagram

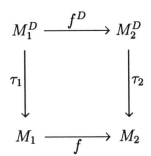

By Proposition 3, (f^D, f) is a morphism from the tangent bundle on M_1 to the tangent bundle on M_2. We then have trivially :

Proposition 5. *The function associating (f^D, f) to f is a functor of the category Mic of microlinear objects into the category Fib of vector bundles.*

We denote by $Fib(M)$ the category of vector bundles over a microlinear object M. This is the sub-category of Fib whose morphisms are the (φ, id_M).

Let V a microlinear R-module ; the projection $\pi_1 : M \times V \to M$ is a vector bundle. We say that this is the trivial bundle of basis M and of fibre V. We say then that M is parallelizable if its tangent bundle $\tau : M^D \to M$ is isomorphic in the category $Fib(M)$, to a trivial bundle $\pi_1 : M \times V \to M$. Note that V is isomorphic to $T_m M$ and is thus an Euclidean R-module. As a very easy example let us prove the following result.

Proposition 6. *If V is a Euclidean microlinear R-module, then V is parallelizable.*

Proof: Let us consider the function $\varphi : V \times V \to V^D$ defined by $\varphi(\underline{v}_0, \underline{v})(d) = \underline{v}_0 + d \cdot \underline{v}$. As V is Euclidean, φ is bijective. It is also immediate that the triangle

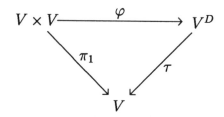

commutes. It remains to see that for every $\underline{v}_0$ in V the mapping $\varphi_{\underline{v}_0} : V \to T_{\underline{v}_0} V$ is linear. But it is obviously homogeneous :

$$
\begin{aligned}
\varphi_{\underline{v}_0}(\lambda \cdot \underline{v})(d) &= \underline{v}_0 + d \cdot (\lambda \cdot \underline{v}) = \underline{v}_0 + (d \cdot \lambda) \cdot \underline{v} \\
&= \varphi_{\underline{v}_0}(v)(\lambda \cdot d) = (\lambda \cdot \varphi_{\underline{v}_0}(v))(d)
\end{aligned}
$$

and thus linear because $T_{\underline{v}_0} V$ is Euclidean (cf. 1.2, Proposition 10). ∎

3.2   Vector fields

3.2.1   The R-modules of vector fields

Let $E \xrightarrow{\xi} M$ be a vector bundle. A section of ξ is a mapping $\sigma : M \to E$ such that $\xi(\sigma(m)) = m$ for all m in M. We write $\Gamma(\xi)$ fot the "set" of sections of ξ .

Proposition 1. *The set $\Gamma(\xi)$ is equipped with a microlinear R-module structure. If ξ is a Euclidean vector bundle, $\Gamma(\xi)$ is a Euclidean R-module.*

Proof: The R-module structure is immediately achieved through the R-module structures of the fibres ξ_m. As E and M are microlinear, E^M and M^M are microlinear (2.3, Proposition 1). Now $\Gamma(\xi)$ is the equalizer of the mappings

$$E^M \underset{c}{\overset{\varphi}{\rightrightarrows}} M^M$$

defined by $\varphi(\sigma) = \xi \circ \sigma$ and $c(\sigma) = id_M$ (i.e. $\Gamma(\xi) = \{\sigma \in E^M | \xi \circ \sigma = id_M\}$). Therefore it is a microlinear object. Finally, if ξ is an Euclidean vector bundle and if $u : D \to \Gamma(\xi)$, we have for all m in M, the existence of a unique element $v(m)$ in ξ_m such that

$$u(d)(m) = u(0)(m) + d \cdot v(m)$$

and so $\Gamma(\xi)$ is Euclidean. ∎

Proposition 2. *$\Gamma(\xi)$ is a module over the ring R^M.*

Proof: For $f : M \to R$ and $\sigma \in \Gamma(\xi)$ we obviously put

$$(f \cdot \sigma)(m) = f(m) \cdot \sigma(m).$$

∎

Definition 1. *A vector field on M is a section of the tangent bundle $\tau : M^D \to M$. The set $\Gamma(\tau)$ of these vector fields is generally denoted by $\mathcal{X}(M)$.*

They are three equivalent ways to see a vector field :
1) It is a mapping
$$X : M \to M^D$$

such that $X(m)(0) = m$. This is the usual concept of vector fields ; at each point m we give a tangent vector at m.

2)It is also a mapping

$$X : M \times D \to M$$

such that $X(m,0) = m$. This corresponds rather to the classical idea of infinitesimal flow or of dynamical micro-systems. The classical additivity of dynamical systems is recuperated infinitesimaly in the following way :

Proposition 3. *If $X : M \times D \to M$ is a vector field we have for every $(d, d') \in D(2)$*

$$X(m, d + d') = X(X(m, d), d').$$

Proof: The two sides of the equality to be proved coincide for every m on the two axes of $D(2)$ with the tangent vector X_m. The equality, thus, results of the microlinearity of M (cfr. 2.3, Proposition 2). ∎

3) Finally, we can see a vector field as a mapping

$$X : D \to M^M$$

such that $X(0) = id_M$. From this point of view a vector field is seen as a tangent vector to the space M^M of transformations of M at the identity. The image X_d of an element d of D is called an infinitesimal transformation of X. Each infinitesimal transformation of X is actually bijective.

Proposition 4. *If $X : D \to M^M$ is a vector field, we have that for each d in D, X_d and X_{-d} are inverse to each other.*

Proof: By Proposition 3, we have

$$X(X(m, d), -d) = X(m, 0) = m.$$

∎

Let us denote by $Iso(M)$ the set of elements of M^M that are invertible. We have :

Proposition 5. *The group $Iso(M)$ is microlinear. The R-module $\mathcal{X}(M)$ of vector fields on M is isomorphic to the tangent space to $Iso(M)$ at its unit element id_M.*

Proof: The set $Iso(M)$ is in bijection with the set

$$\{(f,g) \in M^M \times M^M \mid g \circ f = id_M \text{ and } f \circ g = id_M\}.$$

This set is a finite limit of microlinears. Hence it is microlinear and $Iso(M)$ is microlinear.

The bijection between $\mathcal{X}(M)$ and $T_{id_M}(Iso(M))$ follows from Proposition 4. The fact that this bijection is linear follows then from its homogeneity which is immediate. ∎

It could be interesting to have an explicit description of the sum of two vector fields in terms of infinitesimal tranformations.

Proposition 6. *Let X and Y be two vector fields. The mapping*

$$\lambda : D(2) \rightarrow Iso(M)$$

which coincides on the axes of $D(2)$ with X and Y is given by

$$\lambda(d, d') = X_d \circ Y_{d'} = Y_{d'} \circ X_d.$$

So

$$(X + Y)_d = X_d \circ Y_d = Y_d \circ X_d$$

Proof: The value of the three expressions is X_d at $(d, 0)$, and $Y_{d'}$ at $(0, d')$. Thus the uniqueness clause of the microlinearity condition gives the result. ∎

3.2.2   Lie algebra of vector fields

Let X and Y be two vector fields on M. We define the commutator of associated infinitesimal transformations as being the mapping

$$D \times D \xrightarrow{\tau} M^M$$

given by

$$\tau(d_1, d_2) = Y_{-d_2} \circ X_{-d_1} \circ Y_{d_2} \circ X_{d_1}.$$

We can suggest the way $\tau(d_1, d_2)$ operates at the point m of M with the following figure

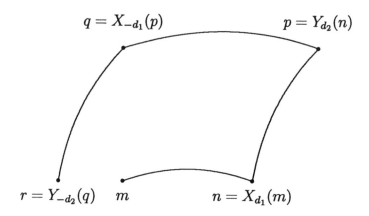

Figure 1

where $r = \tau(d_1, d_2)(m)$. We follow the integral microcurve of the field X during the time d_1 starting from m and we arrive at n. Then, from n on, we follow the integral microcurve of the field Y during the time d_2. After that, we follow that of X during the time $-d_1$ and finally that of Y during the time $-d_2$. There is no reason to come back exactly to the departure spot.

However the function τ is constant on the axes of $D \times D$

$$
\begin{aligned}
\tau(d_1, 0)(m) &= (Y_0 \circ X_{-d_1} \circ Y_0 \circ X_{d_1})(m) \\
&= (X_{-d_1} \circ X_{d_1})(m) \\
&= m \\
\tau(0, d_2)(m) &= (Y_{-d_2} \circ X_0 \circ Y_{d_2} \circ X_0)(m) \\
&= (Y_{-d_2} \circ Y_{d_2})(m) \\
&= m.
\end{aligned}
$$

Now M^M is microlinear and thus satisfies Wraith axiom W (2.3, Proposition 3). The function τ factorizes thus in a unique way through the multiplication. There is one and only one mapping

$$t : D \to M^M$$

such that $\tau(d_1, d_2) = t(d_1 \cdot d_2)$ for all (d_1, d_2) in $D \times D$.

In short, the lack of commutativity of infinitesimal transformations associated with X and Y is measured with a new vector field on M that we call the Lie bracket of X and Y and we denote by $[X, Y]$. Let us record then :

Definition 2. *The Lie bracket* $[X, Y]$ *of vector fields* X *and* Y *is the unique vector field*

$$[X, Y] : D \to M^M$$

satisfying

$$[X, Y](d_1 \cdot d_2) = Y_{-d_2} \circ X_{-d_1} \circ Y_{d_2} \circ X_{d_1}.$$

Let L be an R-module. Let us recall that L is a *Lie R-algebra* if L is equipped with an operation

$$L \times L \to L : (x, y) \rightsquigarrow [x, y]$$

which is bilinear, antisymmetric and satisfies the Jacobi-identity :

$$[[x, y], z] + [[y, z], x] + [[z, x], y] = 0.$$

Proposition 7. *The Lie bracket provides the R-module $\mathcal{X}(M)$ of vector fields with an R-Lie-Algebra structure.*

Proof:

a) Let us first show the antisymmetry

$$[X, Y] = -[Y, X].$$

It suffices to establish the equality of associated infinitesimal tranformations for $d_1 \cdot d_2$ (for whatever d_1 and d_2 in D). As, by virtue of Proposition 4, $(X_d)^{-1} = X_{-d}$ for every field X, we have :

$$
\begin{aligned}
[X, Y]_{d_1 \cdot d_2} &= [X, Y]_{-((-d_1) \cdot d_2)} \\
&= ([X, Y]_{(-d_1) \cdot d_2})^{-1} \\
&= (Y_{-d_2} \circ X_{d_1} \circ Y_{d_2} \circ X_{-d_1})^{-1} \\
&= X_{d_1} \circ Y_{-d_2} \circ X_{-d_1} \circ Y_{d_2} \\
&= [Y, X]_{d_2 \cdot (-d_1)} \\
&= (-[Y, X])_{d_1 \cdot d_2},
\end{aligned}
$$

and the antisymmetry is established.

b) As $\mathcal{X}(M)$ is Euclidean, in order to establish the linearity of $[X,Y]$ at X it suffices to verify its homogeneity

$$[\lambda X, Y] = \lambda[X, Y]$$

which is immediate as the two members have the same value at $d_1 \cdot d_2$.

c) It remains to check the Jacobi identity

$$[[X, Y], Z] + [[Y, Z], X] + [[Z, X], Y] = 0.$$

For this it suffices to verify that the value of the lefthand side is id_M at $d_1 \cdot d_2 \cdot d_3$ for every $(d_1, d_2, d_3) \in D \times D \times D$.

By definition of commutators we have

$$[[X,Y], Z]_{d_1 d_2 d_3}$$
$$= Z_{-d_3} \circ X_{-d_1} \circ Y_{-d_2} \circ X_{d_1} \circ Y_{d_2} \circ Z_{d_3} \circ Y_{-d_2} \circ X_{-d_1} \circ Y_{d_2} \circ X_{d_1}$$

and we obtain in the same way, through circular permutations, $[[Y, Z], X]_{d_1 d_2 d_3}$ and $[[Z, X], Y]_{d_1 d_2 d_3}$. The value of the lefthand side of Jacobi identity in $d_1 d_2 d_3$ is, by Proposition 6, the composite of these three elements :

$$Z_{-d_3} \circ X_{-d_1} \circ Y_{-d_2} \circ X_{d_1} \circ Y_{d_2} \circ Z_{d_3} \circ Y_{-d_2} \circ X_{-d_1} \circ Y_{d_2} \circ X_{d_1}$$
$$\circ X_{-d_1} \circ Y_{-d_2} \circ Z_{-d_3} \circ Y_{d_2} \circ Z_{d_3} \circ X_{d_1} \circ Z_{-d_3} \circ Y_{-d_2} \circ Z_{d_3} \circ Y_{d_2} \quad (1)$$
$$\circ Y_{-d_2} \circ Z_{-d_3} \circ X_{-d_1} \circ Z_{d_3} \circ X_{d_1} \circ Y_{d_2} \circ X_{-d_1} \circ Z_{-d_3} \circ X_{d_1} \circ Z_{d_3}.$$

Let us introduce

$$
\begin{array}{lllll}
Z_{-d_3} \circ Z_{d_3} & \text{between} & Y_{-d_2} & \text{and} & X_{-d_1} \\
X_{d_1} \circ X_{-d_1} & \text{between} & Z_{-d_3} & \text{and} & Y_{d_2} \\
X_{-d_1} \circ X_{d_1} & \text{between} & Z_{-d_3} & \text{and} & Y_{-d_2} \\
Y_{d_2} \circ Y_{-d_2} & \text{between} & X_{-d_1} & \text{and} & Z_{d_3} \\
Y_{-d_2} \circ Y_{d_2} & \text{between} & X_{-d_1} & \text{and} & Z_{-d_3}
\end{array}
$$

we get

$$(1) = Z_{-d_3} \circ X_{-d_1} \circ Y_{-d_2} \circ X_{d_1} \circ [Z, Y]_{d_2 d_3} \circ [Z, X]_{d_1 d_3}$$
$$\circ X_{-d_1} \circ Y_{d_2} \circ [X, Z]_{d_1 d_3} \circ [X, Y]_{d_1 d_2} \circ Y_{-d_2}$$
$$\circ Z_{d_3} \circ [Y, X]_{d_1 d_2} \circ Y_{d_2} \circ Z_{-d_3} \circ X_{d_1} \circ Z_{d_3}.$$

By Proposition 6, if $(d, d') \in D(2)$ and if U and V are two vector fields, we have $U_d \circ V_{d'} = V_{d'} \circ U_d$. As in the previous expression $(d_1 d_2, d_1 d_3)$, $(d_1 d_2, d_2)$, $(d_1 d_2, -d_1)$, $(d_1 d_2, d_2 d_3)$ are in $D(2)$, we can pass $[X, Y]_{d_1 d_2}$ to the left of $[Z, Y]_{d_2 d_3}$. As

$$X_{-d_1} \circ Y_{-d_2} \circ X_{d_1} \circ [X, Y]_{d_1 d_2} = Y_{-d_2},$$

we get

$$
\begin{aligned}
(1) = {} & Z_{-d_3} \circ Y_{-d_2} \circ [Z, Y]_{d_2 d_3} \circ [Z, X]_{d_1 d_3} \circ X_{-d_1} \circ Y_{d_2} \circ [X, Z]_{d_1 d_3} \\
& \circ Y_{-d_2} \circ Z_{d_3} \circ [Y, X]_{d_1 d_2} \circ Y_{d_2} \circ Z_{-d_3} \circ X_{d_1} \circ Z_{d_3}.
\end{aligned}
$$

Let us permute $[Y, X]_{d_1 d_2}$ with Y_{d_2}, observe that

$$Y_{-d_2} \circ Z_{d_3} \circ Y_{d_2} = [Y, Z]_{d_2 d_3} \circ Z_{d_3},$$

and introduce $X_{-d_1} \circ X_{d_1}$ at the end. We get

$$
\begin{aligned}
(1) = {} & Z_{-d_3} \circ Y_{-d_2} \circ [Z, Y]_{d_2 d_3} \circ [Z, X]_{d_1 d_3} \circ X_{-d_1} \circ Y_{d_2} \circ [X, Z]_{d_1 d_3} \\
& \circ [Y, Z]_{d_2 d_3} \circ Z_{d_3} \circ [Y, X]_{d_1 d_2} \circ [Z, X]_{d_1 d_3} \circ X_{d_1}.
\end{aligned}
$$

As $[X, Z]_{d_1 d_3}$ commutes with the three factors that follow, it cancels with $[Z, X]_{d_1 d_3}$. As $Z_{-d_3} \circ Y_{-d_2} \circ [Z, Y]_{d_2 d_3} = Y_{-d_2} \circ Z_{-d_3}$, we obtain :

$$
\begin{aligned}
(1) = {} & Y_{-d_2} \circ Z_{-d_3} \circ [Z, X]_{d_1 d_3} \circ X_{-d_1} \circ Y_{d_2} \\
& \circ [Y, Z]_{d_2 d_3} \circ Z_{d_3} \circ [Y, X]_{d_1 d_2} \circ X_{d_1}.
\end{aligned}
$$

As

$$Z_{-d_3} \circ [Z, X]_{d_1 d_3} \circ X_{-d_1} = X_{-d_1} \circ Z_{-d_3}$$

and

$$Y_{d_2} \circ [Y, Z]_{d_2 d_3} \circ Z_{d_3} = Z_{d_3} \circ Y_{d_2},$$

we have

$$(1) = Y_{-d_2} \circ X_{-d_1} \circ Z_{-d_3} \circ Z_{d_3} \circ Y_{d_2} \circ [Y, X]_{d_1 d_2} \circ X_{d_1}.$$

As

$$Y_{d_2} \circ [Y, X]_{d_1 d_2} \circ X_{d_1} = X_{d_1} \circ Y_{d_2}$$

we have

$$(1) \quad = \quad Y_{-d_2} \circ X_{-d_1} \circ X_{d_1} \circ Y_{d_2}$$
$$= \quad id_M.$$

and the proposition is thus proved.

∎

3.3   Derivations

3.3.1   Directional derivatives

Let M be a microlinear object and X a vector field on M. We intend to associate to X a derivation of the algebra R^M of functions from M to R (Recall that a *derivation* of an R-algebra A is a linear mapping $D : A \to A$ such that $D(f \cdot g) = D(f) \cdot g + f \cdot D(g)$ for every f and g in A).

Let $f : M \to R$. Let us consider for every m in M the function from D to R given by $f(X(m,d))$ for every d in D. It is written in a unique way as a first degree function : the coefficient of d is determined if X, f and m are fixed. We denote it by $L_X(f)(m)$.

Definition 1. *Let $f : M \to R$. The Lie derivative of f in the direction of X is the function*

$$L_X(f) : M \to R$$

defined by

$$f(X(m,d)) = f(m) + d \cdot L_X(f)(m).$$

Instead of $L_X(f)$ we often write $X(f)$.

Proposition 1. *The mapping*

$$L_X : R^M \to R^M$$

is a derivation.

Proof: We proceed in the same easy way as with functions from R to R (cf. 1.2, Proposition 1). For instance, the derivation rule can be obtained by :

$$(f \cdot g)(X(m,d))$$
$$= f(X(m,d)) \cdot g(X(m,d))$$
$$= (f(m) + d \cdot L_X(f)(m)) \cdot (g(m) + d \cdot L_X(g)(m))$$
$$= f(m) \cdot g(m) + d[L_X(f)(m) \cdot g(m) + f(m) \cdot L_X(g)(m)].$$

Hence $L_X(f \cdot g) = L_X(f) \cdot g + f \cdot L_X(g)$. ∎

Let us now study the properties of the *Lie operator*

$$L : \mathcal{X}(M) \to Der(R^M).$$

Let us first observe that $Der(R^M)$ is a Lie algebra for the bracket operation defined by commutator :

$$[D_1, D_2](f) = D_1(D_2(f)) - D_2(D_1(f)).$$

Proposition 2. *The Lie operator*

$$L : \mathcal{X}(M) \to Der(R^M)$$

is a homomorphism of Lie R-algebras.

Proof: Involving as it does, Euclidean R-modules, in order to establish the linearity of L, it suffices to establish its homogeneity (1.2, Proposition 10). This homogeneity is immediate :

$$f((\alpha X)(m,d)) = f(X(m, \alpha \cdot d))$$
$$= f(m) + \alpha \cdot d \cdot L_X(f)(m)$$

and thus $L_{\alpha X}(f) = \alpha \cdot L_X(f)$ for every f.

Let us show now that

$$L_{[X,Y]}(f) = L_X(L_Y(f)) - L_Y(L_X(f)).$$

Let $m \in M$. To compute $f([X,Y](m,d))$ it suffices to do it when $d = d_1 \cdot d_2$. First we have

$$f([X,Y](m, d_1 \cdot d_2)) = f(m) + d_1 \cdot d_2 \cdot L_{[X,Y]}(f)(m). \tag{1}$$

On the other hand, thanks to the Lie-bracket definition :

$$f([X,Y](m,d_1 \cdot d_2)) = f(Y_{-d_2}(X_{-d_1}(Y_{d_2}(X_{d_1}(m))))).$$

In order to simplify notations, we put

$$n = X_{d_1}(m), \; p = Y_{d_2}(n), \; q = X_{-d_1}(p)$$

and $r = Y_{-d_2}(q) = [X,Y]_{d_1 \cdot d_2}(m)$. By applying several times the definition of the Lie derivative, we get :

$$
\begin{aligned}
f(r) = {} & f(q) - d_2 L_Y(f)(q) \\
= {} & f(p) - d_1 L_X(f)(p) - d_2 L_Y(f)(p) + d_1 d_2 L_X(L_Y(f))(p) \\
= {} & f(n) + d_2 L_Y(f)(n) - d_1 L_X(f)(n) \\
& - d_1 d_2 L_Y(L_X(f))(n) - d_2 L_Y(f)(n) \\
& + d_1 d_2 L_X(L_Y(f))(n) \\
= {} & f(m) + d_1 L_X(f)(m) + d_2 L_Y(f)(m) \\
& + d_1 d_2 L_X(L_Y(f))(m) - d_1 L_X(f)(m) \\
& - d_1 d_2 L_Y(L_X(f))(m) - d_2 L_Y(f)(m) \\
& - d_1 d_2 L_X(L_Y(f))(m) + d_1 d_2 L_X(L_Y(f))(m) \\
= {} & f(m) + d_1 d_2 [L_X, L_Y](f)(m). \qquad\qquad (2)
\end{aligned}
$$

By (1) and (2), we find

$$L_{[X,Y]}(f)(m) = [L_X, L_Y](f)(m)$$

for every m and f.   ∎

The R^M-module structure is also preserved.

Proposition 3. *The Lie operator is R^M-linear.*

Proof: Let $X \in \mathcal{X}(M)$ and $\varphi \in R^M$. We have for every $f \in R^M$,

$$
\begin{aligned}
f((\varphi \cdot X)(m,d)) &= f(X(m, \varphi(m) \cdot d)) \\
&= f(m) + \varphi(m) \cdot d \cdot L_X(f)(m)
\end{aligned}
$$

hence $L_{\varphi \cdot X} = \varphi \cdot L_X$.   ∎

Let us also indicate the following classical property, linking the Lie bracket and the Lie derivative :

Proposition 4. Let X and Y be two vector fields and $\varphi : M \to R$. We have :

$$[X, \varphi \cdot Y] = \varphi \cdot [X, Y] + L_X(\varphi) \cdot Y.$$

Proof: It is enough to see that the two sides coincide at $d_1 \cdot d_2$ for every $(d_1, d_2) \in D \times D$. Let us first examine the lefthand side :

$$[X, \varphi \cdot Y]_{d_1 \cdot d_2}(m) = (\varphi \cdot Y)_{-d_2}(X_{-d_1}((\varphi Y)_{d_2}(X_{d_1}(m)))).$$

Put

$$
\begin{aligned}
p &= X_{d_1}(m) \\
q &= (\varphi Y)_{d_2}(X_{d_1}(m)) \\
&= Y_{d_2 \cdot \varphi(p)}(p) \\
r &= X_{-d_1}(q).
\end{aligned}
$$

We have

$$
\begin{aligned}
[X, \varphi \cdot Y]_{d_1 \cdot d_2}(m) &= (\varphi \cdot Y)_{-d_2}(r) \\
&= Y_{-d_2\varphi(r)}(r) \qquad\qquad (1)
\end{aligned}
$$

Notice that $d_2\varphi(r) = d_2\varphi(m)$:

$$
\begin{aligned}
d_2\varphi(r) &= d_2(\varphi(q) - d_1 L_X(\varphi)(q)) \\
&= d_2\varphi(p) - d_1 d_2 L_X(\varphi)(p) \\
&= d_2(\varphi(m) + d_1 L_X(\varphi)(m)) - d_1 d_2 L_X(\varphi)(m) \\
&= d_2\varphi(m).
\end{aligned}
$$

Thus (1) can be written

$$
\begin{aligned}
[X, \varphi \cdot Y]_{d_1 \cdot d_2}(m) &= Y_{-d_2\varphi(m)}(r) \\
&= Y_{-d_2\varphi(m)}(X_{-d_1}(Y_{d_2\varphi(p)}(X_{d_1}(m))))
\end{aligned}
$$

Now

$$
\begin{aligned}
Y_{d_2\varphi(p)} &= Y_{d_2\varphi(m) + d_1 d_2 L_X(\varphi)(m)} \\
&= Y_{d_2\varphi(m)} \circ Y_{d_1 d_2 L_X(\varphi)(m)}
\end{aligned}
$$

(cf. 3.2, Proposition 3). Thus we have

$$[X, \varphi \cdot Y]_{d_1 \cdot d_2}(m) = Y_{-d_2\varphi(m)}(X_{-d_1}(Y_{d_2\varphi(m)}(Y_{d_1d_2L_X(\varphi)(m)}(X_{d_1}(m))))).$$

Now, as $(d_1d_2L_X(\varphi)(m), d_1) \in D(2)$, the last two infinitesimal transformations commute and we get

$$[X, \varphi \cdot Y \ ]_{d_1 \cdot d_2}(m)$$
$$= Y_{-d_2 \cdot \varphi(m)}(X_{-d_1}(Y_{d_2 \cdot \varphi(m)}(X_{d_1}(Y_{d_1d_2L_X(\varphi)(m)}(m)))))$$
$$= [X, Y]_{d_1 \cdot d_2 \cdot \varphi(m)}(Y_{d_1d_2L_X(\varphi)(m)}(m)). \tag{2}$$

Let us now examine the righthand side of the equality that we have to prove. We have :

$$(\varphi \cdot [X, Y] + L_X(\varphi) \cdot Y)_{d_1 \cdot d_2}(m)$$
$$= (\varphi \cdot [X, Y])_{d_1 \cdot d_2}((L_X(\varphi) \cdot Y)_{d_1 \cdot d_2}(m))$$
$$= (\varphi \cdot [X, Y])_{d_1 \cdot d_2}(Y_{d_1 \cdot d_2 \cdot L_X(\varphi)(m)}(m))$$
$$= [X, Y]_{d_1 \cdot d_2 \cdot \varphi(p)}(p)$$

where $p = Y_{d_1 \cdot d_2 L_X(\varphi)(m)}(m)$. We immediately have $d_1 \cdot d_2 \cdot \varphi(p) = d_1 \cdot d_2 \cdot \varphi(m)$ and thus we get

$$(\varphi \cdot [X, Y] + L_X(\varphi) \cdot Y)_{d_1 \cdot d_2}(m)$$
$$= [X, Y]_{d_1 \cdot d_2 \cdot \varphi(m)}(Y_{d_1 \cdot d_2 \cdot L_X(\varphi)(m)}(m))$$

which is (2). The proposition is proved. ∎

3.3.2   Reflexive objects

There is no need *a priori* for the Lie operator

$$L : \mathcal{X}(m) \rightarrow \text{Der}(R^M)$$

to be bijective. However in classical differential geometry if M is a C^∞ paracompact manifold, the Lie operator is a bijection between the Lie algebra of C^∞ vector fields on M and the Lie algebra of derivations of the algebra of C^∞ functions from M to $\mathbb{R}$. Is there an analogy in synthetic differential geometry of this notion of "paracompact manifold" ?

The analogous notion could well be that of reflexive object introduced by A. Kock [31].

Definition 2. *An object M is a reflexive object if the evaluation*

$$M \xrightarrow{ev} alg(R^M, R)$$

is bijective, where $alg(R^M, R)$ denote the set of R-algebra homomorphisms.

(The evaluation associates to $m \in M$ the homomorphism of R-algebras $ev_m : R^M \rightarrow R$ whose value at $\varphi \in R^M$ is $\varphi(m)$). By analogy with the functional analysis terminology, a reflexive object could also be called replete. Note also that in classical differential geometry, a C^∞ manifold M is paracompact if and only if the evaluation

$$M \xrightarrow{ev} alg(C^\infty(M, \mathbb{R}), \mathbb{R})$$

is bijective.

Let us first observe the following easy result.

Proposition 5. *If M is reflexive, M is microlinear.*

Proof: By axiom (K-W) R is microlinear, thus R^M is microlinear (2.3 Prop. 1) and so the set $alg(R^M, R)$ of R-algebra homomorphisms is microlinear. If the evaluation is bijective, M is thus microlinear. ∎

We must check now that this concept of reflexive objects does achieve the desired end :

Proposition 6. *If M is reflexive, then the Lie operator $L : \mathcal{X}(M) \rightarrow Der(R^M)$ is bijective.*

Proof: Let us remember that $L_X : R^M \rightarrow R^M$ is characterized by

$$f(X(m, d)) = f(m) + d \cdot L_X(f)(m).$$

Let us construct $\Lambda : Der(R^M) \rightarrow \mathcal{X}(M)$. Let $U \in Der(R^M)$. We define

$$X_U : M \times D \rightarrow alg(R^M, R)$$

by putting

$$X_U(m, d)(f) = f(m) + d \cdot U(f)(m).$$

We first observe that $X_U(m,d)$ is actually an homomorphism of R-algebras : we have for instance

$$\begin{aligned}
X_U(m,d)(f \cdot g) &= (f \cdot g)(m) + d \cdot U(f \cdot g)(m) \\
&= f(m) \cdot g(m) + d \cdot (U(f)(m) \cdot g(m) \\
&\quad + f(m) \cdot U(g)(m))
\end{aligned}$$

and

$$\begin{aligned}
X_U(m,d)(f) &\cdot X_U(m,d)(g) = \\
&= (f(m) + d \cdot U(f)(m)) \cdot (g(m) + d \cdot U(g)(m)) \\
&= f(m) \cdot g(m) + d \cdot (U(f)(m) \cdot g(m) + f(m) \cdot U(g)(m)) \cdot
\end{aligned}$$

As M is reflexive, the evaluation ev is a bijection from M to $alg(R^M, R)$. We can thus characterize a vector field $\Lambda(U)$ by

$$ev(\Lambda(U)(m,d)) = X_U(m,d).$$

It is actually a vector field because

$$ev(\Lambda(U)(m,0))(f) = X_U(m,0)(f) = f(m)$$

and thus $\Lambda(U)(m,0) = m$.

Let us show that $\Lambda(L_X) = X$. We have

$$\begin{aligned}
ev(\Lambda(L_X)(m,d))(f) &= f(m) + d \cdot L_X(f)(m) \\
&= f(X(m,d)) \\
&= ev(X(m,d))(f)
\end{aligned}$$

hence $\Lambda(L_X) = X$.

Conversely let us show that $L_{\Lambda(U)} = U$. First $L_{\Lambda(U)}(f)$ is characterized by

$$f(\Lambda(U)(m,d)) = f(m) + d \cdot L_{\Lambda(U)}(f)(m). \qquad (1)$$

On the other hand :

$$\begin{aligned}
f(\Lambda(U)(m,d)) &= ev(\Lambda(U)(m,d))(f) \\
&= X_U(m,d)(f) \\
&= f(m) + d \cdot U(f)(m). \qquad (2)
\end{aligned}$$

As (1) and (2) are true for every d, we find $L_{\Lambda(U)}(f) = U(f)$.   ∎

The same can be done at one point m. Let $h = ev_m$ (the evaluation at m). A derivation from R^M to R relative to h is a linear function $D : R^M \to R$ such that

$$D(f \cdot g) = D(f) \cdot h(g) + h(f) \cdot D(g).$$

We denote by Der_h the R-module of these relative derivations. If $t \in T_m M$ we define a relative derivation $\lambda(t) \in Der_h$ by

$$f(t(d)) = f(m) + d \cdot \lambda(t)(f).$$

In the spirit of the last proposition, it is easy to prove :

Proposition 7. *If M is reflexive, $m \in M$ and $h = ev_m$, the operator $\lambda : T_m(M) \to Der_h$ is bijective.*

∎

It might be of interest to consider the abstract situation of a unitary, associative and commutative R-algebra A. A point of A is a homomorphism of R-algebras $h : A \to R$. A vector on A at h is a derivation from A to R relative to h. A field on A is a derivation of A. If M is reflexive, $A = R^M$, this terminology is natural.

All this would be worthless if we could not guarantee the existence of sufficiently many reflexive objects. On this subject, the results are not complete, but we shall dedicate the rest of this section to the description of a class of reflexive objects.

Amazingly, in order to see that R and R^n are reflexive, we shall need the integration axiom or at least the resulting Fermat-Reyes axiom (cf. 1.3.4) :

$$\forall f : R \to R \, \exists! g : R \times R \to R \, \forall x, y \in R$$

$$f(y) - f(x) = (y - x)g(x, y).$$

We can extend this to an analogous proposition with parameters :

Lemma. *Let X be an object and $f : X \times R \to R$. There is one and only one function $g : X \times R \times R \to R$ with $\forall \alpha \in X, \forall x, y \in R$,*

$$f(\alpha, y) - f(\alpha, x) = (y - x)g(\alpha, x, y).$$

Proof: This results by applying the above mentioned proposition to every α of X.   ∎

Proposition 8. *For every n and for every $f : R^n \to R$, there is a $g_i : R^n \times R^n \to R$ $(i = 1, 2, \ldots, n)$ with $\forall \underline{x}, \underline{y} \in R^n$:*

$$f(\underline{y}) - f(\underline{x}) = \sum_{i=1}^{n}(y_i - x_i)g_i(\underline{x}, \underline{y}).$$

Proof: We have

$$
\begin{aligned}
f(\underline{y}) - f(\underline{x}) &= f(y_1, y_2, \ldots, y_n) - f(x_1, y_2, \ldots, y_n) \\
&\quad + f(x_1, y_2, \ldots, y_n) - f(x_1, x_2, \ldots, y_n) \\
&\quad + \ldots \\
&\quad + f(x_1, \ldots, x_{i-1}, y_i, y_{i+1}, \ldots, y_n) \\
&\quad\quad - f(x_1, \ldots, x_{i-1}, x_i, y_{i+1}, \ldots, y_n) \\
&\quad + \ldots \\
&\quad + f(x_1, x_2, \ldots, x_{n-1}, y_n) - f(x_1, x_2, \ldots, x_n).
\end{aligned}
$$

Put $X_i = R^{n-1}$ and let

$$
\begin{aligned}
f_i \; &: \; X_i \times R \to R \\
&: \; (u_1, \ldots, u_{n-1}, x) \rightsquigarrow f(u_1, \ldots, u_{i-1}, x, u_i, \ldots, u_{n-1}).
\end{aligned}
$$

By the lemma, there is $g_i : X_i \times R \times R \to R$ such that

$$
\begin{aligned}
f(u_1, \ldots, u_{i-1}, y, u_i, \ldots, u_{n-1}) &- f(u_1, \ldots, u_{i-1}, x, u_i, \ldots, u_{n-1}) \\
&= (y - x)g_i(u_1, \ldots, u_{n-1}, x, y).
\end{aligned}
$$

By putting

$$
\begin{aligned}
(u_1, \ldots, u_{i-1}) &= (x_1, \ldots, x_{i-1}) \\
(u_i, \ldots, u_{n-1}) &= (y_{i+1}, \ldots, y_n)
\end{aligned}
$$

and considering thus g_i as a function of $\underline{x}$ and $\underline{y}$, we get :

$$f(\underline{y}) - f(\underline{x}) = \sum_{i=1}^{n}(y_i - x_i)g_i(\underline{x}, \underline{y}).$$

■

Proposition 9. *If R satisfies the integration axiom (or more generally the Fermat-Reyes axiom), then R^n is reflexive.*

Proof: Let us define

$$\mu : alg(R^{(R^n)}, R) \to R^n$$

by putting $\mu(U) = (U(\pi_i))_{1 \leq i \leq n}$ and verify that μ is inverse to the evaluation. We first have

$$\mu(ev(\underline{x})) = (ev(\underline{x})(\pi_i))_{1 \leq i \leq n} = \underline{x}.$$

Conversely let us show that $ev(\mu(U))(f) = U(f)$ for every $f : R^n \to R$. Put $\mu(U) = \underline{a}$ (i.e. $a_i = U(\pi_i)$). The previous proposition gives :

$$f(\underline{x}) - f(\underline{a}) = \sum_{i=1}^{n} (x_i - a_i) g_i(\underline{x}, \underline{a}).$$

For a fixed $\underline{a}$, it is an equality between functions of $\underline{x}$. By applying U we obtain

$$
\begin{aligned}
U(f) - f(\underline{a}) &= \sum_{i=1}^{n} (U(\pi_i) - a_i) U(g_i(\underline{x}, \underline{a})) \\
&= 0
\end{aligned}
$$

and so $U(f) = f(\underline{a}) = ev(\mu(U))(f)$ as claimed. ∎

Consider the diagram

$$M \overset{i}{\hookrightarrow} R^n \overset{f}{\longrightarrow} R.$$

Definition 3. *We say that M is a reflexive immersed manifold in R^n if*

a) *M is the kernel of f :*

$$M = \{\underline{x} \in R^n | f(\underline{x}) = 0\}.$$

b) *Every function from M to R extends to a function from R^n to R, in other words $R^i : R^{(R^n)} \to R^M$ is surjective. We say also that M is extensible.*

The reason for saying "immersed manifold", is that in classical differential geometry every paracompact C^∞ manifold M can be immersed in some $\mathbb{R}^n$ so that there exists f, a C^∞ function from $\mathbb{R}^n$ to $\mathbb{R}$ such that M is the kernel of f.

The reason for saying "reflexive" is the following proposition.

Proposition 10. *Let*

$$M \overset{i}{\hookrightarrow} R^n \overset{f}{\longrightarrow} R.$$

be a reflexive immersed manifold of R^n. Then M is reflexive.

Proof: As R^n is reflexive we have the inverse

$$\mu : alg(R^{(R^n)}, R) \to R^n$$

of the evaluation. By composition with the application

$$alg(R^M, R) \overset{\tilde{i}}{\longrightarrow} alg(R^{(R^n)}, R)$$

induced by the inclusion $i : M \to R^n$, we find a mapping

$$\tilde{\mu} : alg(R^M, R) \to R^n.$$

Let us show that $\tilde{\mu}$ factorizes through M. Let $U \in alg(R^M, R)$. We have

$$\mu(\tilde{i}(U)) = (\tilde{i}(U(\pi_k)))_{1 \le k \le n}$$

Let us note $\underline{a} = \mu(\tilde{i}(U))$. For every function $\psi : R^n \to R$ we have

$$\tilde{i}(U)(\psi) = U(\psi \circ i) = \psi(\underline{a}).$$

As $i : M \to R^n$ is a kernel of f , we have for every component f_j $(j = 1, \ldots, p)$:

$$f_j(\underline{a}) = U(f_j \circ i) = 0$$

and thus $f(\underline{a}) = 0$ and $\underline{a} \in M$.

We immediately have for every x in M

$$
\begin{aligned}
\tilde{\mu}(ev(x)) &= \mu(\tilde{i}(ev(x))) \\
&= (\tilde{i}(ev(x)(\pi_k)))_{1 \le k \le n} \\
&= x
\end{aligned}
$$

Let us show that for all $\varphi : M \to R$,

$$ev(\tilde{\mu}(U))(\varphi) = U(\varphi).$$

As M is extensible, there is $\psi : R^n \to R$ with $\psi \circ i = \varphi$. We then have

$$ev(\tilde{\mu}(U))(\varphi) = \varphi(\underline{a}) = \psi(\underline{a}) = U(\psi \circ i) = U(\varphi),$$

which ends the proof.   ■

3.4   Micro-squares

We have seen in section 3.1.1, a tangent vector to M as a kind of micro-segment drawed on M

$$t : D \to M.$$

Naturally we can draw on M micro-objects of higher dimension.
    A *n-microcube* on M is a mapping

$$\gamma : D^n \to M.$$

(Following the terminology of algebraic topology, we could speak of "singular" n-microcube).
    Here, we shall only be concerned with the study of algebraic structure on the set $M^{D \times D}$ of *micro-squares*.
    We can first consider $M^{D \times D}$ as an *iterated tangent bundle*. More precisely, let $i_1 : D \to D \times D : d \rightsquigarrow (0, d)$ and $i_2 : D \to D \times D : d \rightsquigarrow (0, d)$ be the injections of the axes of $D \times D$.

Proposition 1. *The mappings*

$$M^{i_2} : M^{D \times D} \to M^D \text{ and } M^{i_1} : M^{D \times D} \to M^D$$

are Euclidean vector bundles.

Proof: We have the isomorphism

$$\alpha : M^{D \times D} \xrightarrow{\sim} (M^D)^D$$

that associates to $\gamma : D \times D \to M : (d_1, d_2) \rightsquigarrow \gamma(d_1, d_2)$

$$\alpha(\gamma)(d_1)(d_2) = \gamma(d_1, d_2)$$

and the isomorphism

$$\beta : M^{D \times D} \xrightarrow{\sim} (M^D)^D$$

that associates $\beta(\gamma)(d_2)(d_1) = \gamma(d_1, d_2)$ to γ. By composing M^{i_2} (resp. M^{i_1}) with α^{-1} (resp. β^{-1}) we get the tangent bundle to M^D. The proposition is thus established. ∎

We fix some notation. If $t : D \to M : d_2 \rightsquigarrow t(d_2)$ is in M^D, the fibre of $M^{i_2} : M^{D \times D} \to M^D$ over t is denoted by $M_{1,t}$ (or sometimes $M^{i_2}(t)$). The addition is denoted by $\underset{1}{+}$ and the scalar multiplication $\underset{1}{\cdot}$. If γ_1 and γ_2 lie over t (i.e. $\gamma_1(0, d_2) = t(d_2) = \gamma_2(0, d_2)$), we get

$$(\gamma_1 \underset{1}{+} \gamma_2)(d_1, d_2) = \ell_1(d_1, d_1, d_2)$$

where $\ell_1 : D(2) \times D \to M : (d, d', d_2) \rightsquigarrow \ell_1(d, d', d_2)$ is characterized by

$$\ell_1(d, 0, d_2) = \gamma_1(d, d_2) \text{ and } \ell_1(0, d', d_2) = \gamma_2(d', d_2).$$

The scalar multiplication is given by

$$(\alpha \underset{1}{\cdot} \gamma)(d_1, d_2) = \gamma(\alpha \cdot d_1, d_2).$$

We adopt analogous notations (with the index 2) in the fibres of M^{i_1}.

A simple permutation of variables obviously transforms one vector bundle into the other. More precisely, let

$$\Sigma : M^{D \times D} \to M^{D \times D}$$

be the mapping induced by the symmetry

$$\sigma : D \times D \to D \times D$$

$$(d_1, d_2) \rightsquigarrow (d_2, d_1).$$

Proposition 2. *The mapping Σ is a morphism of vector bundles from M^{i_2} to M^{i_1} and from M^{i_1} to M^{i_2}.*

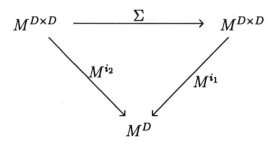

Proof: That Σ sends fibres to fibres follows immediately from

$$\sigma \circ i_1 = i_2 \quad \text{and} \quad \sigma \circ i_2 = i_1.$$

That Σ is linear on the fibre follows from its homogeneity, which is trivial. For instance :

$$
\begin{aligned}
(\Sigma(\alpha \underset{1}{\cdot} \gamma))(d_1, d_2) &= (\alpha \underset{1}{\cdot} \gamma)(\sigma(d_1, d_2)) \\
&= (\alpha \underset{1}{\cdot} \gamma)(d_2, d_1) = \gamma(\alpha \cdot d_2, d_1) \\
&= \gamma(\sigma(d_1, \alpha \cdot d_2)) = \Sigma(\gamma)(d_1, \alpha \cdot d_2) \\
&= (\alpha \underset{2}{\cdot} \Sigma(\gamma))(d_1, d_2).
\end{aligned}
$$

∎

It is also interesting to observe that the two structures possess the following permutability properties.

Proposition 3.

(a) For every micro-square γ and every scalar α and β, we have

$$\beta \underset{2}{\cdot} (\alpha \underset{1}{\cdot} \gamma) = \alpha \underset{1}{\cdot} (\beta \underset{2}{\cdot} \gamma)$$

(b) let γ_1, γ_2, γ_1' et γ_2' be four micro-squares with

$$\gamma_1 \circ i_2 = \gamma_2 \circ i_2; \ \gamma_1 \circ i_1 = \gamma_1' \circ i_1; \ \gamma_1' \circ i_2 = \gamma_2' \circ i_2; \ \gamma_2 \circ i_1 = \gamma_2' \circ i_1;$$

we have

$$(\gamma_1 \underset{1}{+} \gamma_2) \underset{2}{+} (\gamma_1' \underset{1}{+} \gamma_2') = (\gamma_1 \underset{2}{+} \gamma_1') \underset{1}{+} (\gamma_2 \underset{2}{+} \gamma_2'). \ (*)$$

We also have

$$
\begin{aligned}
\alpha \underset{2}{\cdot} (\gamma_1 \underset{1}{+} \gamma_2) &= (\alpha \underset{2}{\cdot} \gamma_1) \underset{1}{+} (\alpha \underset{2}{\cdot} \gamma_2) \\
\text{and} \quad \alpha \underset{1}{\cdot} (\gamma_1 \underset{2}{+} \gamma_1') &= (\alpha \underset{1}{\cdot} \gamma_1) \underset{2}{+} (\alpha \underset{1}{\cdot} \gamma_1').
\end{aligned}
$$

Proof: Part (a) is trivial, we prove part (b). We verify easily, computing by testing with R, that the following diagram is a quasi-colimit :

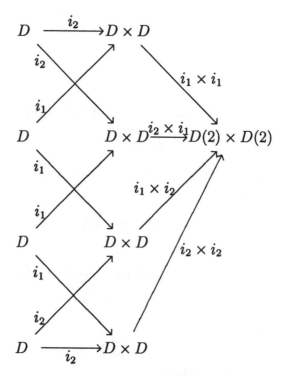

As M is micro-linear, we have a unique function

$$\lambda : D(2) \times D(2) \to M$$

such that

$$
\begin{aligned}
\lambda(d,0,\delta,0) &= \gamma_1(d,\delta) \\
\lambda(0,d',\delta,0) &= \gamma_2(d',\delta) \\
\lambda(d,0,0,\delta') &= \gamma_1'(d,\delta') \\
\lambda(0,d',0,\delta') &= \gamma_2'(d',\delta').
\end{aligned}
$$

But then we have

$$
\begin{aligned}
\lambda(d,d,\delta,0) &= (\gamma_1 \underset{1}{+} \gamma_2)(d,\delta) \\
\lambda(d,d,0,\delta) &= (\gamma_1' \underset{1}{+} \gamma_2')(d,\delta)
\end{aligned}
$$

and thus

$$\lambda(d,d,\delta,\delta) = [(\gamma_1 \underset{1}{+} \gamma_2) \underset{2}{+} (\gamma_1' \underset{1}{+} \gamma_2')](d,\delta).$$

We also have

$$\lambda(d,0,\delta,\delta) = (\gamma_1 \underset{2}{+} \gamma_1')(d,\delta)$$
$$\lambda(0,d,\delta,\delta) = (\gamma_2 \underset{2}{+} \gamma_2')(d,\delta)$$

thus

$$\lambda(d,d,\delta,\delta) = [(\gamma_1 \underset{2}{+} \gamma_1') \underset{1}{+} (\gamma_2 \underset{2}{+} \gamma_2')](d,\delta)$$

and the proposition is proved. ∎

Note that if we put

$$t = \gamma_1 \circ i_2 = \gamma_2 \circ i_2 \, ; \, t_1 = \gamma_1 \circ i_1 = \gamma_1' \circ i_1$$
$$t' = \gamma_1' \circ i_2 = \gamma_2' \circ i_2 \, ; \, t_2 = \gamma_2 \circ i_1 = \gamma_2' \circ i_1$$

the composite with i_2 of the two sides of (*) is $t+t'$ and their composite with i_1 is $t_1 + t_2$.

This description does not exhaust the structure of $M^{D\times D}$. Let $i : D(2) \to D \times D$ be the injection. We shall study

$$M^i : M^{D\times D} \to M^{D(2)}.$$

Remember first the notion of affine space. Let V be an R-module and A a set. We say that A is equipped with an affine space structure over V if we specify the operations

$$V \times A \xrightarrow{\dotplus} A$$

and

$$A \times A \xrightarrow{\,\dot{-}\,} V$$

satisfying the following identities for every γ, γ_1, γ_2 in A and every t, t_1, t_2 in V :

(a) $(\gamma_2 \dot{-} \gamma_1) \dotplus \gamma_1 = \gamma_2$

(b) $(t \dotplus \gamma) \dot{-} \gamma = t$

(c) $0 \dotplus \gamma = \gamma$

(d) $(t_1 \dotplus t_2) + \gamma = t_1 \dotplus (t_2 \dotplus \gamma)$.

Consider a fibre of the mapping

$$M^i : M^{D \times D} \to M^{D(2)},$$

in other words, for $\tau : D(2) \to M$ fixed, the set M^i_τ of micro-squares

$$\gamma : D \times D \to M$$

whose restriction to $D(2)$ is τ. Let us denote $x = \tau(0,0)$ the contact point in M. We intend to see that M^i_τ has an affine space structure on the R-module $T_x M$.

a)   We shall first define the "difference", often celled *strong differ-ence*,

$$\dot{-} : M^i_\tau \times M^i_\tau \to T_x M.$$

We have the following lemma.

Lemma. *The diagram*

$$
\begin{array}{ccc}
D(2) & \xrightarrow{\;\;i\;\;} & D \times D \\
\Big\downarrow{\scriptstyle i} & & \Big\downarrow{\scriptstyle \psi} \\
D \times D & \xrightarrow{\;\;\varphi\;\;} & (D \times D) \vee D
\end{array}
$$

where $(D \times D) \vee D = \{(d_1, d_2, e) \mid d_1^2 = d_2^2 = e^2 = d_1 e = d_2 e = 0\}$ *and* $\varphi(d_1, d_2) = (d_1, d_2, 0)$ *and* $\psi(d_1, d_2) = (d_1, d_2, d_1, d_2)$ *is a quasi-colimit.*

Proof: It is a cone made up of small objects. By (2.2, Prop. 9), it suffices thus to verify that R perceives the diagram as a colimit. Let thus $\gamma_1 : D \times D \to R$ and $\gamma_2 : D \times D \to R$ coincide on $D(2)$. We can write

$$
\begin{aligned}
\gamma_1(d_1, d_2) &= a + b_1 d_1 + b_2 d_2 + c_1 d_1 d_2 \\
\gamma_2(d_1, d_2) &= a + b_1 d_1 + b_2 d_2 + c_2 d_1 d_2.
\end{aligned}
$$

Let us define $f : (D \times D) \vee D \rightarrow R$ by

$$f(d_1, d_2, e) = a + b_1 d_1 + b_2 d_2 + c_1 d_1 d_2 + (c_2 - c_1)e.$$

We have $f(d_1, d_2, 0) = \gamma_1(d_1, d_2)$ and $f(d_1, d_2, d_1 \cdot d_2) = \gamma_2(d_1, d_2)$ and f is the only function from $(D \times D) \vee D$ to R satisfying these identities. ∎

Let γ_1 and $\gamma_2 : D \times D \rightarrow M$ with $\gamma_1|_{D(2)} = \gamma_2|_{D(2)} = \tau$. We define

$$\gamma_2 \dot{-} \gamma_1 : D \rightarrow M$$

with $(\gamma_2 \dot{-} \gamma_1)(d) = f(0, 0, d)$ where $f : (D \times D) \vee D \rightarrow M$ is the only function such that

$$\begin{aligned} f(\varphi(d_1, d_2)) &= f(d_1, d_2, 0) = \gamma_1(d_1, d_2) \\ f(\psi(d_1, d_2)) &= f(d_1, d_2, d_1 d_2) = \gamma_2(d_1, d_2). \end{aligned}$$

b)   Next, we define the "addition"

$$\dot{+} : T_x M \times M_\tau^i \rightarrow M_\tau^i.$$

We have first the following easy lemma (left as exercise in 2.2.1).

Lemma. *The diagram*

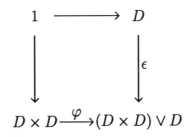

where $\varphi(d_1, d_2) = (d_1, d_2, 0)$ and $\epsilon(d) = (0, 0, d)$, is a quasi-colimit. ∎

Let $t : D \rightarrow M$ and $\gamma : D \times D \rightarrow M$ with $t(0) = \gamma(0, 0)$. There is a unique mapping

$$u : (D \times D) \vee D \rightarrow M$$

such that $u \circ \varphi = \gamma$ and $u \circ \epsilon = t$. We define

$$t \dot{+} \gamma : D \times D \rightarrow M$$

with $(t \dot{+} \gamma)(d_1, d_2) = u(\psi(d_1, d_2)) = u(d_1, d_2, d_1 \cdot d_2).$

Proposition 4. *The operations* $\dot{-}$ *and* $\dot{+}$ *equip* M_r^i *with an affine space structure on* $T_x M$.

Proof:

a) Let us show that

$$(\gamma_2 \dot{-} \gamma_1) \dot{+} \gamma_1 = \gamma_2.$$

By definition of $\dot{+}$, the value of the lefthand side at (d_1, d_2) is $u(d_1, d_2, d_1 \cdot d_2)$ where u is characterized by $u(d_1, d_2, 0) = \gamma_1(d_1, d_2)$ and $u(0, 0, d) = (\gamma_2 \dot{-} \gamma_1)(d)$. But the definition of $\dot{-}$ gives $(\gamma_2 \dot{-} \gamma_1)(d) = \ell(0, 0, d)$ where ℓ satisfies $\ell(d_1, d_2, 0) = \gamma_1(d_1, d_2)$ and $\ell(d_1, d_2, d_1 \cdot d_2) = \gamma_2(d_1, d_2)$. Hence we have $u = \ell$ and thus $u(d_1, d_2, d_1 \cdot d_2) = \gamma_2(d_1, d_2)$.

b) We verify in the same way that

$$(t \dot{+} \gamma) \dot{-} \gamma = t$$

and, trivially,

c) $0 \dot{+} \gamma = \gamma$.

d) Let us show that

$$(t_1 + t_2) \dot{-} \gamma = t_1 \dot{-} (t_2 \dot{-} \gamma).$$

We first verify that the following diagram is a quasi colimit.

$$
\begin{array}{ccc}
D \times D & \xrightarrow{\;\varphi\;} & (D \times D) \vee D \\
\downarrow{\scriptstyle\psi} & & \downarrow{\scriptstyle\zeta_2} \\
(D \times D) \vee D & \xrightarrow{\;\zeta_1\;} & (D \times D) \vee D(2)
\end{array}
$$

where

$$
\begin{aligned}
(D \times D) \vee D(2) &= \{(d_1, d_2, e_1, e_2) | 0 = d_1^2 = d_2^2 \\
&\quad = e_1^2 = e_2^2 = d_i e_j = e_1 e_2\} \\
\zeta_1(d_1, d_2, e) &= (d_1, d_2, e, 0) \\
\zeta_2(d_1, d_2, e) &= (d_1, d_2, d_1 \cdot d_2, e).
\end{aligned}
$$

It suffices indeed to test it with functions with values in R.

We then consider

$$
u : (D \times D) \vee D \to M
$$

characterized by $u(0, 0, e) = t_2(e)$ and $u(d_1, d_2, 0) = \gamma(d_1, d_2)$, and also

$$
v : (D \times D) \vee D \to M
$$

characterized by $v(0, 0, e) = t_1(e)$ and

$$
v(d_1, d_2, 0) = u(d_1, d_2, d_1.d_2).
$$

We have thus $u \circ \psi = v \circ \varphi$ and there exists a unique mapping

$$
w : (D \times D) \times D(2) \to M
$$

such that $w \circ \zeta_1 = u$ and $w \circ \zeta_2 = v$.

Having obviously

$$
[t_1 \dot{+} (t_2 \dot{+} \gamma)](d_1, d_2) = v(d_1, d_2, d_1 \cdot d_2),
$$

it suffices to verify that we also have

$$
[(t_1 + t_2) \dot{+} \gamma](d_1, d_2) = v(d_1, d_2, d_1 \cdot d_2).
$$

Put $z(d_1, d_2, e) = w(d_1, d_2, e, e)$. We have

$$
z(d_1, d_2, 0) = w(d_1, d_2, 0, 0) = u(d_1, d_2, 0) = \gamma(d_1, d_2)
$$

and

$$
z(0, 0, e) = w(0, 0, e, e) = (t_1 + t_2)(e)
$$

because

$$w(0,0,e,0) = t_2(e) \text{ and } w(0,0,0,e) = t_1(e).$$

Hence, we get

$$
\begin{aligned}
((t_1 + t_2)\dot{+}\gamma)(d_1, d_2) &= z(d_1, d_2, d_1 \cdot d_2) \\
&= w(d_1, d_2, d_1 \cdot d_2, d_1 \cdot d_2) \\
&= v(d_1, d_2, d_1 \cdot d_2),
\end{aligned}
$$

as claimed.

∎

As for the scalar multiplication, we easily get the following properties.

Proposition 5. *Let γ_1 and γ_2 be two microsquares coinciding on $D(2)$, and let $\alpha \in R$. We have*

$$
\begin{aligned}
\alpha \cdot (\gamma_2 \dot{-} \gamma_1) &= (\alpha \underset{1}{;} \gamma_2) \dot{-} (\alpha \underset{1}{;} \gamma_1) \\
&= (\alpha \underset{2}{;} \gamma_2) \dot{-} (\alpha \underset{2}{;} \gamma_2).
\end{aligned}
$$

In the same way for every microsquare γ and every tangent vector at the same point t we have

$$
\begin{aligned}
\alpha \underset{1}{;} (t \dot{+} \gamma) &= (\alpha \cdot t) \dot{+} (\alpha \underset{1}{;} \gamma) \\
\alpha \underset{2}{;} (t \dot{+} \gamma) &= (\alpha \cdot t) \dot{+} (\alpha \underset{2}{;} \gamma).
\end{aligned}
$$

∎

We point out also the compatibility with Σ (cf. Proposition 2).

Proposition 6. *If γ_1 and γ_2 coincide on $D(2)$, we have*

$$\Sigma(\gamma_2) \dot{-} \Sigma(\gamma_1) = \gamma_2 \dot{-} \gamma_1.$$

The action $\dot{+}$ satisfies also

$$\Sigma(t \dot{+} \gamma) = t \dot{+} \Sigma(\gamma).$$

∎

It is possible to express the strong difference through the two fibre bundle structures. Let γ_1 and γ_2 be two microsquares coinciding on $D(2)$. Let $\tilde{t}_1$ and $\tilde{t}_2$ be two microsquares squashed on the axis, i.e.

$$\tilde{t}_1(d_1, d_2) = \gamma_1(d_1, 0) = \gamma_2(d_1, 0)$$

and

$$\tilde{t}_2(d_1, d_2) = \gamma_1(0, d_2) = \gamma_2(0, d_2).$$

Proposition 7. *We have two identities, each of which characterizes the strong difference :*

$$
\begin{aligned}
(\gamma_2 \overset{\cdot}{-} \gamma_1)(d_1 \cdot d_2) &= ((\gamma_2 \underset{1}{-} \gamma_1) \underset{2}{-} \tilde{t}_2)(d_1, d_2) \\
&= ((\gamma_2 \underset{2}{-} \gamma_1) \underset{1}{-} \tilde{t}_1)(d_1, d_2).
\end{aligned}
$$

Proof: By symmetry, it suffices to prove the first equality. Let

$$\zeta : (D \times D) \vee D \to M$$

be the mapping characterized by

$$\zeta(d_1, d_2, 0) = \gamma_1(d_1, d_2) \text{ and } \zeta(d_1, d_2, d_1 \cdot d_2) = \gamma_2(d_1, d_2).$$

We have thus $\zeta(0, 0, d_1 \cdot d_2) = (\gamma_2 \overset{\cdot}{-} \gamma_1)(d_1 \cdot d_2)$. We define $\ell_1 : (D \vee D) \times D \to M$ by

$$\ell_1(d_1, d_1', d_2) = \zeta(d_1 - d_1', d_2, d_1 \cdot d_2)$$

(which makes sense because $(d_1 - d_1', d_2, d_1 \cdot d_2) \in (D \times D) \vee D$). We have

$$
\begin{aligned}
\ell_1(0, d_1', d_2) &= \zeta(-d_1', d_2, 0) = \gamma_1(-d_1', d_2) \\
\ell_1(d_1, 0, d_2) &= \zeta(d_1, d_2, d_1 \cdot d_2) = \gamma_2(d_1, d_2)
\end{aligned}
$$

and thus

$$\zeta(0, d_2, d_1 \cdot d_2) = \ell_1(d_1, d_1, d_2) = (\gamma_2 \underset{1}{-} \gamma_1)(d_1, d_2).$$

We consider then $\ell_2 : D \times (D \vee D) \to M$ given by

$$\ell_2(d_1, d_2, d_2') = \zeta(0, d_2 - d_2', d_1 \cdot d_2).$$

We have

$$
\begin{aligned}
\ell_2(d_1, 0, d_2') &= \zeta(0, -d_2', 0) = -\tilde{t}_2(d_2') \\
\ell_2(d_1, d_2, 0) &= \zeta(0, d_2, d_1 \cdot d_2) = (\gamma_2 \underset{1}{-} \gamma_1)(d_1, d_2).
\end{aligned}
$$

Thus we have

$$
\begin{aligned}
\zeta(0, 0, d_1 \cdot d_2) &= \ell_2(d_1, d_2, d_2) \\
&= ((\gamma_2 \underset{1}{-} \gamma_1) \underset{2}{-} \tilde{t}_2)(d_1, d_2),
\end{aligned}
$$

which yields the required identity.

By (2.2, Proposition 5), the strong difference is then uniquely determined. ∎

As an application of this affine-space structure of the set $M^{D \times D}$ of microsquares, let us point out that the Lie bracket of two vector fields can be described as the difference of two microsquares. Let X and Y be two vector fields. We look upon them as tangent vectors to M^M at id_M (cf. 3.2)

$$
\begin{aligned}
X &: D \to M^M : d \rightsquigarrow X_d \\
Y &: D \to M^M : d \rightsquigarrow Y_d.
\end{aligned}
$$

We define the microsquare

$$
Y \star X : D \times D \to M^M
$$

by

$$
(Y \star X)(d_1, d_2) = Y_{d_2} \circ X_{d_1}.
$$

Proposition 8. *We have*

$$
[X, Y] = Y \star X \,\dot{-}\, \Sigma(X \star Y).
$$

Proof: Notice that

$$
\begin{aligned}
(Y \star X)(d_1, d_2) &= Y_{d_2} \circ X_{d_1} \\
\Sigma(X \star Y)(d_1, d_2) &= X_{d_1} \circ Y_{d_2}
\end{aligned}
$$

thus on $D(2)$ these two microsquares coincide and the difference is well-defined. Let $\ell : (D \times D) \vee D \to M^M$ be given by

$$\ell(d_1, d_2, e) = X_{d_1} \circ [X, Y]_e \circ Y_{d_2}.$$

We have

$$
\begin{aligned}
\ell(d_1, d_2, 0) &= X_{d_1} \circ Y_{d_2} = \Sigma(X \star Y)(d_1, d_2) \\
\ell(d_1, d_2, d_1.d_2) &= X_{d_1} \circ [X, Y]_{d_1 \cdot d_2} \circ Y_{d_2} \\
&= X_{d_1} \circ (X_{-d_1} \circ Y_{d_2} \circ X_{d_1} \circ Y_{-d_2}) \circ Y_{d_2} \\
&= Y_{d_2} \circ X_{d_1} = (Y \star X)(d_1, d_2)
\end{aligned}
$$

Thus the difference is

$$(Y \star X \dot{-} \Sigma(X \star Y))(e) = \ell(0, 0, e) = [X, Y]_e.$$

∎

Finally let us point out that we have a surjective mapping $M^{D \times D} \to M$ that evaluates every microsquare at its center $(0, 0)$. A section

$$\sigma : M \to M^{D \times D}$$

of this surjection is called a *distribution of dimension 2* on M (it is the bidimensional similar of a vector field).

We may say that a field

$$X : M \to M^D$$

belongs to the distribution of dimension 2, σ, if for all x of M there exists a mapping from D to $D \times D$ making commutative the diagram

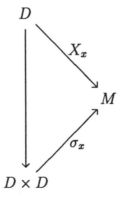

We say that a distribution σ of dimension 2 is involutive if, for every X and Y belonging to σ, $[X, Y]$ belongs also to σ.

3.5   Commented bibliography

1) The idea that a tangent vector at a suitable object M be conceived as a map from D to M, is part of the basis of synthetic differential geometry, and is found, explicitly, in the seminal lectures of Lawvere [44]. It comes from Grothendieck's differential calculus on algebraic manifolds, and also from the concept of jet of Ehresmann [17].

2) The concept of vector field is also basic. Reyes and Wraith [54] (1978), with suitable infinitesimal linearity hypotheses, give a Lie algebra structure to the space of vector fields. Kock explores the derivation operation in [30] (1979). Some complements, as Koszul's formula, are in Lavendhomme [40] (1981).

3) The vector bundle structures on the space of microsquares appear in Kock and Reyes in [36](1979). The affine structure on the space of microsquares is explained by Kock and Lavendhomme in [35] (1984).

Chapter 4

Differential forms

4.1 Differential forms with values in an R-module

4.1.1 Singular definition

Let M be a microlinear object and E be a microlinear Euclidean R-module. A differential n-form on M with value in E will be a function that associates to every n-microcube on M an element of E and this in an n-linear and alternated way, in a sense that will be specified.

Let $M^{(D^n)}$ be the set of n-microcubes in M. We generalize as follows the fibre bundle structures described in section 3.4 in the case of microsquares. Let

$$i_k : D^{n-1} \to D^n$$

be the function inserting 0 at k^{th} position. We easily have the generalization of (3.4, Prop. 1).

Proposition 1. *The mappings*

$$M^{i_k} : M^{(D^n)} \to M^{(D^{n-1})}$$

are Euclidean fibre bundles. ∎

For instance the multiplication by a scalar α in a fibre of M^{i_k} is given by

$$(\alpha \underset{k}{\cdot} \gamma)(d_1, \ldots, d_n) = \gamma(d_1, \ldots, \alpha \cdot d_k, \ldots, d_n).$$

If σ is a permutation of the numbers $1, 2, \ldots, n$, we define

$$\Sigma : M^{(D^n)} \to M^{(D^n)}$$

by

$$\Sigma(\gamma)(d_1, \ldots, d_n) = \gamma(d_{\sigma(1)}, \ldots, d_{\sigma(n)}).$$

We can then define :

Definition 1. *A differential n-form on M with value in E is a mapping*

$$\omega : M^{(D^n)} \to E$$

which is :

a) *n-homogeneous, in the sense that for every k, $(1 \leq k \leq n)$,*

$$\omega(\alpha \underset{k}{\cdot} \gamma) = \alpha \cdot \omega(\gamma)$$

for every γ in $M^{(D^n)}$ and every α in R.

b) *alternated, in the sense that for every permutation σ of the numbers $1, 2, \ldots, n$,*

$$\omega(\Sigma(\gamma)) = \epsilon_\sigma \omega(\gamma)$$

for every γ of $M^{(D^n)}$ (where ϵ_σ is the parity of σ with value $+1$ or -1 depending to whether σ is even or odd).

We denote by $\Omega_n(M; E)$ the set of differential n-forms on M with values in E. In case of ambiguity we shall sometimes specify that the so called differential forms are defined in a "singular" sense or that they are *singular differential forms*.

Notice that (1.2.4, Prop. 10) the n-homogeneity implies that, for every k, the restriction of ω to a fibre of $M^{i_k} : M^{(D^n)} \to M^{(D^{n-1})}$ is R-linear.

As $\Omega_n(M; E)$ is a part of $E^{(M^{(D^n)})}$ described by simple conditions (of n-homogeneity and alternance), we have trivially the following proposition :

Proposition 2. *$\Omega_n(M; E)$ is a microlinear Euclidean R-module.* ∎

If $\omega : M^{(D^n)} \to E$ is a differential form and $\varphi : M \to R$ is a function, we define $\varphi \cdot \omega : M^{(D^n)} \to E$ by

$$(\varphi \cdot \omega)(\gamma) = \varphi(\gamma(\underline{0})) \cdot \omega(\gamma)$$

and we have :

Proposition 3. $\Omega_n(M; E)$ *is a module over the algebra* R^M. ∎

Let us indicate the (contravariant) functoriality of $\Omega_n(M; E)$ with respect to M. Let $f : M \to N$ (with M and N microlinear). We define

$$f^* : \Omega_n(N; E) \to \Omega_n(M; E)$$

by $(f^*\omega)(\gamma) = \omega(f \circ \gamma)$ for every ω in $\Omega_n(N; E)$ and γ in $M^{(D^n)}$. Then we have :

Proposition 4. *Let M and N be microlinear and $f : M \to N$. For every $\omega \in \Omega_n(N; E)$, $f^*\omega$ is in $\Omega_n(M; E)$. Furthermore*

$$f^* : \Omega_n(N; E) \to \Omega_n(M; E)$$

is R-linear. The assignment $f \rightsquigarrow f^*$ is (contravariant) functorial.

Proof: All the verifications are easy. As an example let us verify that $f^*\omega$ is n-homogeneous.

$$
\begin{aligned}
(f^*\omega)(\alpha \underset{i}{\cdot} \gamma) &= \omega(f \circ (\alpha \underset{i}{\cdot} \gamma)) = \omega(\alpha \underset{i}{\cdot} (f \circ \gamma)) \\
&= \alpha \cdot \omega(f \circ \gamma) = \alpha \cdot (f^*\omega)(\gamma).
\end{aligned}
$$

∎

Finally let us point out that if we define

$$f^* : R^N \to R^M$$

by composition $(f^*(\varphi) = \varphi \circ f)$ we have the following compatibility condition :

$$f^*(\varphi \cdot \omega) = f^*\varphi \cdot f^*\omega$$

because

$$
\begin{aligned}
f^*(\varphi \cdot \omega)(\gamma) = (\varphi \cdot \omega)(f \circ \gamma) &= \varphi(f(\gamma(\underline{0}))) \cdot \omega(f \circ \gamma) \\
&= (f^*\varphi \cdot f^*\omega)(\gamma).
\end{aligned}
$$

4.1.2   Classical differential forms

Had we copied more rigorously the notion of classical differential forms we would have been led to a less wide definition of differential forms. Let x be a point of M and $L_a^n(T_x M, E)$ the R-module of n-linear alternated mappings from $T_x M \times \ldots \times T_x M$ to E. A classical differential n-form $\tilde{\omega}$ can be considered as the datum at each x of M of an element ω_x of $L_a^n(T_x M, E)$. More generally, we can consider

$$M^D \underset{M}{\times} M^D \underset{M}{\times} \cdots \underset{M}{\times} M^D$$

defined by

$$\{(t_1, \ldots, t_n) \mid t_1, \ldots, t_n \in M^D, t_1(0) = \ldots = t_n(0)\}.$$

A *classical differential form* is a mapping from this set to E which is n-linear and alternated. As M is microlinear, it is n-infinitesimally linear (2.3, Prop. 2) and thus we have an isomorphism

$$M^{D(n)} \xrightarrow{\sim} M^D \underset{M}{\times} M^D \underset{M}{\times} \cdots \underset{M}{\times} M^D$$

(that associates to each $\tau : D(n) \to M$ the n restrictions to the axis of $D(n)$). Then a classical differential form is an n-homogeneous and alternated mapping :

$$\tilde{\omega} : M^{D(n)} \to E.$$

Let us consider the injection $i : D(n) \to D^n$. To every classical differential form $\tilde{\omega}$, we associate the mapping $\omega = \tilde{\omega} \circ M^i$ and we trivially verify that ω is still n-homogeneous and alternated :

Proposition 5. *The function*

$$\tilde{\omega} \rightsquigarrow \omega = \tilde{\omega} \circ M^i$$

associates to every classical differential form a singular differential form.

■

In some cases the "singular" differential forms and "classical" differential forms correspond to each other bijectively. It is of course the case for forms of degree one, where the two notions coincide. Let us explore the case of forms of degree two.

Let $i : D(2) \to D \times D$ be the canonical injection. If V is a Euclidean R-module, the mapping

$$V^i : V^{D \times D} \to V^{D(2)}$$

has a "canonical" section : it is the function

$$\nabla : V^{D(2)} \to V^{D \times D}$$

that to

$$\tilde{\gamma} : D(2) \to V : (d_1, d_2) \rightsquigarrow a_0 + a_1 d_1 + a_2 d_2$$

associates

$$\gamma : D \times D \to V : (d_1, d_2) \rightsquigarrow a_0 + a_1 d_1 + a_2 d_2.$$

This mapping ∇ is linear for both structures of fibre bundle :

$$\nabla(\alpha \underset{1}{\cdot} \gamma) = \alpha \underset{1}{\cdot} \nabla(\gamma)$$

$$\nabla(\alpha \underset{2}{\cdot} \gamma) = \alpha \underset{2}{\cdot} \nabla(\gamma)$$

and possesses the symmetry property :

$$\nabla(\gamma \circ \sigma) = \nabla(\gamma) \circ \sigma$$

(where σ is the permutation $(d_1, d_2) \rightsquigarrow (d_2, d_1)$ in $D(2)$ or in $D \times D$). We could say that ∇ describes the flat connection of the Euclidean R-module V, but the analysis of these concepts supposes a connection-notion which we shall deal with in the next chapter. However let us give prematurely the following definition.

Definition 2. *A symmetric connection on an object M is a section*

$$\nabla : M^{D(2)} \to M^{D \times D}$$

of M^i which is :

a) linear for both structures of fibre bundle :

$$\nabla(\alpha \underset{1}{\cdot} \gamma) = \alpha \underset{1}{\cdot} \nabla(\gamma)$$

$$\nabla(\alpha \underset{2}{\cdot} \gamma) = \alpha \underset{2}{\cdot} \nabla(\gamma)$$

b) *symmetric (we also say traditionally "without torsion") i.e. :*

$$\nabla(\gamma \circ \sigma) = \nabla(\gamma) \circ \sigma.$$

Our only aim in defining this concept here is to be able to formulate more easily the following proposition.

Proposition 6. *If M is equipped with a symmetric connection, the function $w \rightsquigarrow w = \tilde{w} \circ M^i$ (where $i : D(2) \to D \times D$ is the canonical injection) is a bijection from the set of the classical differential 2-forms to the set of singular differential 2-forms.*

   Proof: Let

$$\nabla : M^{D(2)} \to M^{D \times D}$$

be the symmetrical connection given on M.

a) The two linearity conditions and the symmetry condition assure that the function $w \rightsquigarrow \tilde{w} = w \circ \nabla$ does transform a singular differential form into a classical differential form. As ∇ is a section of M^i we have

$$\tilde{w} \circ M^i \circ \nabla = \tilde{w}$$

b) Let w be a differential (singular) form of degree 2. We are going to show that if γ_1 and γ_2 are two microsquares of M coinciding on the axis (i.e. $\gamma_1 \circ i = \gamma_2 \circ i$) then $w(\gamma_1) = w(\gamma_2)$. If we remember the affine structure described in (3.4, Prop. 4), we can consider the strong difference

$$\gamma_2 \overset{\cdot}{-} \gamma_1$$

and thus the microsquare α given by

$$\alpha(d_1, d_2) = (\gamma_2 \overset{\cdot}{-} \gamma_1)(d_1 \cdot d_2).$$

By (3.4, Prop. 7) we have

$$\alpha = (\gamma_2 \underset{i}{\overset{\cdot}{-}} \gamma_1) \underset{2}{\overset{\cdot}{-}} \tilde{t}_2$$

where $\tilde{t}_2(d_1, d_2)$ is the square squashed onto the second axis : $\tilde{t}_2(d_1, d_2) = \gamma_1(0, d_2) = \gamma_2(0, d_2)$. Of course,

$$w(\tilde{t}_2) = w(0 \underset{i}{\cdot} \tilde{t}_2) = 0 \cdot w(\tilde{t}_2) = 0$$

and by the symmetry of α, $\omega(\alpha) = -\omega(\alpha)$, thus $\omega(\alpha) = 0$. By 2-linearity of ω, we have $\omega(\gamma_2 \underset{1}{-} \gamma_1) = 0$ and by 1-linearity of ω, we have indeed $\omega(\gamma_1) = \omega(\gamma_2)$.

c) It follows from remark b) above that $\omega \circ \nabla \circ M^i = \omega$ since for every microsquare γ, $\nabla(M^i(\gamma))$ and γ have the same restrictions to $D(2)$. We thus have the announced bijection.

■

Using conditions on the existence of connections of suitable higher orders, we could extend proposition 6 to higher degrees. We simply state without proof, what happens in degree 3.

Let j_1 be the injection of $D(3)$ into $D[3]$ where

$$D[3] = \{(d_1, d_2, d_3) \in D^3 \mid d_1 \cdot d_2 \cdot d_3 = 0\}$$

and let j_2 the injection of $D[3]$ into D^3. A symmetrical 3-connection is a couple (∇_1, ∇_2) where ∇_k is a section of $M^{j_k}(k = 1, 2)$ and where

a)

$$\nabla_r(\alpha \cdot_l \gamma_r) = \alpha \cdot_l \nabla_r(\gamma_r)$$

for $r = 1, 2; l = 1, 2, 3; \alpha \in R, \gamma_1$ in $M^{d(3)}$ and γ_2 in $M^{D[3]}$.

b) For σ a permutation of (d_1, d_2, d_3) we have

$$\nabla_r(\gamma_r \circ \sigma) = \nabla(\gamma_r)$$

for $r = 1, 2, \gamma_1 \in M^{D(3)}$ and $\gamma_2 \in M^{D[3]}$.

If M is equipped with a symmetrical 3-connection, we obtain a bijection from the set of classical 3-forms to the set of singular 3-formes associating to the classical 3-form $\tilde{\omega}$ the singular 3-form $\omega = \tilde{\omega} \circ M^{j_1} \circ M^{j_2}$. This bijection has as inverse the function associating to ω the classical 3-form $\tilde{\omega} = \omega \circ \nabla_2 \circ \nabla_1$.

4.2   The exterior differential

4.2.1   Infinitesimal n-chains

A disadvantage of n-microcubes $\gamma : D^n \to M$ is that they do not have a distinct edge. But we can arbitrarily confer one on them by giving an element $\underline{e}$ in D^n. We shall say that a couple $(\gamma, \underline{e})$, where $\gamma : D^n \to M$ and $\underline{e} \in D^n$, is a *marked microcube*. The set of marked n-microcubes is thus $M^{D^n} \times D^n$. We give the following definition.

Definition 1. *The R-module of* infinitesimal n-chains *is the free R-module generated by the marked n-microcubes.*

This R-module is refered to by $C_n(M)$. An infinitesimal n-chain is thus a formal linear combination, with coefficients in R,

$$\sum_{i=1}^{r} \alpha_i \cdot (\gamma_i, \underline{e}_i)$$

where $\alpha_i \in R$, $\gamma_i \in M^{D^n}$, $\underline{e}_i \in D^n$. By definition of free R-modules the linear mappings of $C_n(M)$ into a R-module E correspond bijectively to the functions from the set $M^{D^n} \times D^n$ of marked n-microcubes into E.

We are now in position to describe the *boundary operator*

$$\partial : C_{n+1}(M) \to C_n(M).$$

It suffices to describe the infinitesimal n-chain boundary of a marked $(n+1)$-microcube and then to extend the definition by linearity.

Thus let $(\gamma, \underline{e})$ be a marked $(n+1)$-microcube. For every i, $1 \leq i \leq n+1$, we consider the two sides of $(\gamma, \underline{e})$ that are the two marked n-microcubes

$$F_0^i(\gamma, \underline{e}) = (\gamma_0^i; (e_1, \ldots, \hat{e}_i, \ldots, e_{n+1}))$$

and

$$F_1^i(\gamma, \underline{e}) = (\gamma_1^i; (e_1, \ldots, \hat{e}_i, \ldots, e_{n+1}))$$

where :

$$\gamma_0^i : D^n \to M$$

and

$$\gamma_1^i : D^n \to M$$

are the n-microcubes given by

$$\begin{aligned} \gamma_0^i(d_1,\ldots,d_n) &= \gamma(d_1,\ldots,d_{i-1},0,d_i,\ldots,d_n) \\ \gamma_1^i(d_1,\ldots,d_n) &= \gamma(d_1,\ldots,d_{i-1},e_i,d_i,\ldots,d_n). \end{aligned}$$

We then put

$$\partial(\gamma,\underline{e}) = \sum_{i=1}^{n+1}(-1)^i F_0^i(\gamma,\underline{e}) - \sum_{i=1}^{n+1}(-1)^i F_1^i(\gamma,\underline{e}).$$

For instance for $n=1$, $\gamma: D^2 \to M$, $(e_1,e_2) \in D^2$ we have :

$$\partial(\gamma,(e_1,e_2)) = F_0^2(\gamma,\underline{e}) - F_0^1(\gamma,\underline{e}) + F_1^1(\gamma,\underline{e}) - F_1^2(\gamma,\underline{e})$$

$$= (\gamma(-,0),e_1) - (\gamma(0,-),e_2) + (\gamma(e_1,-),e_2) - (\gamma(-,e_2),e_1)$$

which is represented by the following drawing :

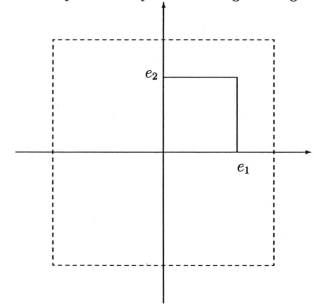

Figure 2

Proposition 1. We have $\partial \circ \partial = 0$.

Proof: It suffices to show that for every marked $(n+1)$-microcube $(\gamma,\underline{e})$, we have $\partial(\partial(\gamma,\underline{e})) = 0$. We have

$$
\begin{aligned}
\partial(\partial(\gamma,\underline{e})) &= \partial(\sum_{i=1}^{n+1}(-1)^i F_0^i(\gamma,\underline{e}) - \sum_{i=1}^{n+1}(-1)^i F_1^i(\gamma,\underline{e})) \\
&= \sum_{i=1}^{n+1}(-1)^i \partial(F_0^i(\gamma,\underline{e})) - \sum_{i=1}^{n+1}(-1)^i \partial(F_1^i(\gamma,\underline{e})) \\
&= \sum_{i=1}^{n+1}(-1)^i [\sum_{j=1}^{n}(-1)^j F_0^j(F_0^i(\gamma,\underline{e})) - \sum_{j=1}^{n}(-1)^j F_1^j(F_0^i(\gamma,\underline{e}))] \\
&- \sum_{i=1}^{n+1}(-1)^i [\sum_{j=1}^{n}(-1)^j F_0^j(F_1^i(\gamma,\underline{e})) - \sum_{j=1}^{n}(-1)^j F_1^j(F_1^i(\gamma,\underline{e}))].
\end{aligned}
$$

It suffices then to verify that all the terms vanish two-by-two by checking that for $1 \le j < i \le n+1$ and for $\epsilon = 0$ or 1 and $\eta = 0$ or 1, we have

$$
F_\epsilon^j(F_\eta^i(\gamma,\underline{e})) = F_\eta^{i-1}(F_\epsilon^j(\gamma,\underline{e}))
$$

which follows from the definition of the two sides. ∎

4.2.2   Integral of a n-form on an infinitesimal n-chain

Let M again be microlinear and E be a microlinear and Euclidean R-module. Let $\omega : M^{D^n} \to E$ be a differential n-form with value in E. We want to define a linear mapping

$$
\int \omega : \quad C_n(M) \to E.
$$

For that it suffices to define it on the marked n-microcubes :

Definition 2. *The integral of a differential n-form ω on the marked n-microcube $(\gamma,\underline{e})$ is defined by :*

$$
\int_{(\gamma,\underline{e})} \omega = e_1 \cdot e_2 \cdot \ldots \cdot e_n \cdot \omega(\gamma).
$$

Actually this type of microintegration is enough to characterize the n-form ω. More precisely, let

$$\varphi : M^{D^n} \times D^n \to E$$

be a mapping such that

1) for each $\underline{e}$ of D^n, $\varphi(-, \underline{e})$ is n-homogeneous;

2) for each γ of M^{D^n}, $\varphi(\gamma, -)$ is n-homogeneous;

3) for each permutation σ of the numbers $1, 2, \ldots, n$ we have

$$\varphi(\gamma \circ \sigma, \underline{e}) = \epsilon_\sigma \varphi(\gamma, \underline{e} \circ \sigma)$$

(where $\underline{e} \circ \sigma$ is $(e_{\sigma(1)}, e_{\sigma(2)}, \ldots, e_{\sigma(n)})$).

Proposition 2. *If $\varphi : M^{D^n} \times D^n \to E$ verifies the conditions (1), (2) and (3) here above, there is one and only one differential n-form ω such that*

$$\varphi(\gamma, \underline{e}) = \int_{(\gamma, \underline{e})} \omega.$$

Proof: Let us consider the function

$$\tilde{\varphi} : D^n \to E^{(M^{D^n})}$$

$$\underline{e} \rightsquigarrow \varphi(-, \underline{e}).$$

By (2), $\tilde{\varphi}$ vanishes if one of the e_i is zero. But the diagram :

$$D^{n-1} \overset{\underset{\cdots}{\overset{f_1}{\longrightarrow}}}{\underset{\underset{0}{\overset{f_n}{\longrightarrow}}}{\longrightarrow}} D^n \overset{m}{\longrightarrow} D$$

where

$$f_i(d_1, \ldots, d_{n-1}) = (d_1, \ldots, d_{i-1}, 0, d_i, \ldots, d_{n-1})$$

$$m(e_1, \ldots, e_n) = e_1 \cdot e_2 \ldots e_n,$$

s a quasi colimit, in fact, R perceives it as a coequalizer (the verification s similar to the one of (2.2, Prop. 7)).

Since E is microlinear, $E^{(M^{D^n})}$ is also microlinear and thus perceives the previous diagram as a coequalizer. Hence, there is a unique mapping

$$\psi : D \to E^{(M^{D^n})}$$

such that

$$\psi(e_1 \cdot e_2 \cdots \cdots e_n) = \varphi(-, \underline{e}).$$

As E is Euclidean, $E^{(M^{D^n})}$ is Euclidean (1.1, Prop. 4) and as $\psi(0) = 0$, there is one and only only element $w \in E^{(M^{D^n})}$ such that

$$\psi(d) = d \cdot w.$$

Thus, we have a uniquefunction w from M^{D^n} to E such that

$$\varphi(\gamma, \underline{e}) = e_1 \cdot e_2 \cdots \cdots e_n \cdot w(\gamma).$$

As φ satisfies (1) and (3), w is n-homogeneous and alternated. It is a differential n-form and we have

$$\varphi(\gamma, \underline{e}) = \int_{(\gamma, \underline{e})} w.$$

■

By definition, $\int_c w$ is a linear function of $c \in C_n(M)$. But it is also a linear function of $w \in \Omega_n(M; E)$.

4.2.3   The exterior differential

Let w be a differential n-form with value in E. To it we wish to associate a differential $(n+1)$-form . Let $(\gamma, \underline{e})$ be a marked $(n+1)$-microcube. Let us define

$$\varphi(\gamma, \underline{e}) = \int_{(\gamma, \underline{e})} w.$$

We want to verify that φ satisfies the hypotheses (1), (2) and (3) of Proposition 2. Then it will follow from this proposition that φ determines, in a unique way, a differential $(n+1)$-form which we denote by dw and that we call *exterior differential* of w.

Using the notation of Section 4.2.1, we first have :

$$\varphi(\gamma, \underline{e}) = \sum_{i=1}^{n+1}(-1)^i \int_{F_0^i(\gamma, \underline{e})} \omega - \sum_{i=1}^{n+1}(-1)^i \int_{F_1^i(\gamma, \underline{e})} \omega$$

$$= \sum_{i=1}^{n+1}(-1)^i e_1 \cdot e_2 \cdots \cdots \hat{e}_i \cdots \cdots e_{n+1}(\omega(\gamma_0^i) - \omega(\gamma_1^i)).$$

Let $e \in D$ and define the n-microcube $\gamma^i(e)$ by :

$$\gamma^i(e)(d_1, \ldots, d_n) = \gamma(d_1, \ldots, d_{i-1}, e, d_i, \ldots, d_n)$$

let $F^i(e) = \omega(\gamma^i(e)) \in E$. Thus F_i is a function from D to E. As E is Euclidean, it can be written in a unique way as a function of the first degree :

$$F^i(e) = F^i(0) + e \cdot DF^i(0).$$

As $\omega(\gamma_0^i) = F^i(0)$ and $\omega(\gamma_1^i) = F^i(e_i)$, we get :

$$\varphi(\gamma, \underline{e}) = \sum_{i=1}^{n+1}(-1)^i e_1 \cdot e_2 \cdots \cdots \hat{e}_i \cdots \cdots e_{n+1}(-e_i DF^i(0))$$

$$= e_1 \cdot e_2 \cdots \cdots e_{n+1} \cdot \sum_{i=1}^{n+1}(-1)^{i+1} DF^i(0).$$

This expression shows that, for all γ, $\varphi(\gamma, -)$ is an $(n+1)$-homogeneous function of $\underline{e}$. Let us show the $(n+1)$-homogeneity with respect to the variable γ. Let j be a fixed index between 1 and $n+1$ and let $\alpha \in R$. We must prove j-homogeneity :

$$\varphi(\alpha \underset{j}{;} \gamma, \underline{e}) = \alpha \cdot \varphi(\gamma, \underline{e}).$$

To that end we use the initial expression :

$$\varphi(\gamma, \underline{e}) = \sum_{i=1}^{n+1}(-1)^i e_1 e_2 \ldots \hat{e}_i \ldots e_{n+1}(\omega(\gamma_0^i) - \omega(\gamma_1^i)).$$

Let us show j-homogeneity of each term, thus of each expression $\omega(\gamma_0^i) - \omega(\gamma_1^i)$, i.e. $\omega(\gamma^i(0)) - \omega(\gamma^i(e_i))$:

a) for $i > j$, we immediately have

$$
\begin{aligned}
(\alpha \, ; \gamma)^{i}(e)(d_1, \ldots, d_n) &= \gamma(d_1, \ldots, \alpha \cdot d_j, \ldots, d_{i-1}, e, \ldots, d_n) \\
&= (\alpha \, ; \gamma^{i}(e))(d_1, \ldots, d_n)
\end{aligned}
$$

and thus the desired j-homogeneity follows from the j-homogeneity of w.

b) for $i < j$, we have in the same way

$$
(\alpha \, ; \gamma)^{i}(e) = \alpha_{(j \, \dot{-} \, 1)} \gamma^{i}(e)
$$

and j-homogeneity follows from $(j-1)$-homogeneity of w.

c) finally for $i = j$ we have

$$
(\alpha \, ; \gamma)^{i}(e) = \gamma^{i}(\alpha \cdot e)
$$

and thus

$$
\begin{aligned}
w((\alpha \, ; \gamma)^{i}(0)) - w((\alpha \, ; \gamma)^{i}(e_i)) &= w(\gamma^{i}(0)) - w(\gamma^{i}(\alpha \cdot e_i)) \\
&= -\alpha \cdot e_i \cdot DF^{i}(0) \\
&= \alpha \cdot (w(\gamma^{i}(0)) - w(\gamma^{i}(e_i))).
\end{aligned}
$$

It remains to establish the alternating property. Let σ be a permutation of the numbers $1, 2, \ldots, n+1$. We must show that

$$
\varphi(\gamma \circ \sigma, \underline{e}) = \epsilon_{\sigma} \varphi(\gamma, \underline{e} \circ \sigma).
$$

We have

$$
\varphi(\gamma \circ \sigma, \underline{e}) = \sum_{i=1}^{n+1} (-1)^{i} e_1 \ldots \ldots \hat{e}_i \ldots \ldots e_{n+1}
$$

$$
(w((\gamma \circ \sigma)_0^{i}) - w((\gamma \circ \sigma)_1^{i})).
$$

We have

$$
\begin{aligned}
(\gamma \circ \sigma)^{i}(e)(d_1, \ldots, d_n) &= (\gamma \circ \sigma)(d_1, \ldots, d_{i-1}, e, d_i, \ldots, d_n) \\
&= \gamma(d_{\tau(1)}, \ldots, d_{\tau(j-1)}, e, d_{\tau(j)}, \ldots, d_{\tau(n)})
\end{aligned}
$$

where $j = \sigma^{-1}(i)$ and

$$\tau(1) = \sigma(1); \ldots; \tau(j-1) = \sigma(j-1)$$
$$\tau(j) = \sigma(j+1); \ldots; \tau(n) = \sigma(n+1)$$

Notice that $\epsilon_\tau = \epsilon_\sigma \cdot (-1)^{j-i}$. Thus

$$\varphi(\gamma \circ \sigma, \underline{e}) = \sum_{i=1}^{n+1} (-1)^i e_1. \ldots. \hat{e}_i. \ldots .e_{n+1}$$
$$\cdot \epsilon_\sigma (-1)^{j-i} \cdot (\omega(\gamma_0^j) - \omega(\gamma_1^j))$$

where $j = \sigma^{-1}(i)$. We can write

$$\varphi(\gamma \circ \sigma, \underline{e}) = \sum_{j=1}^{n+1} (-1)^j \, \epsilon_\sigma \, e_1. \ldots. e_{\widehat{\sigma(j)}}. \ldots .e_{n+1}(\omega(\gamma_0^j) - \omega(\gamma_1^j))$$

which is

$$\epsilon_\sigma \varphi(\gamma, \underline{e} \circ \sigma).$$

Thus we have shown that Proposition 2 can be applied to $\varphi(\gamma, \underline{e})$:

Proposition 3. *There exists one and only one differential $(n+1)$-form, denoted by $d\omega$, such that*

$$\int_{\partial(\gamma,\underline{e})} \omega = \int_{(\gamma,\underline{e})} d\omega$$

for every marked $(n+1)$-microcube $(\gamma, \underline{e})$. ∎

Definition 3. *The form $d\omega$ is called the differential or exterior differential of ω.*

During the proof we got an explicit expression for

$$\varphi(\gamma, \underline{e}) = d\omega(\gamma)e_1 \ldots e_{n+1}$$

Proposition 4. *We have*

$$d\omega(\gamma) = \sum_{i=1}^{n+1} (-1)^{i+1} DF^i(0)$$

where F^i is given by $F^i(e) = \omega(\gamma^i(e))$ with

$$\gamma^i(e)(d_1, \ldots, d_n) = \gamma(d_1, \ldots, d_{i-1}, e, d_i, \ldots, d_n)$$

∎

Linearity follows immediately from Proposition 3.

Proposition 5. *The differential operator*

$$d : \Omega_n(M; E) \to \Omega_{n+1}(M; E)$$

is R-linear.   ∎

From Propositions 1 and 3 :

Proposition 6. *We have :*

$$d \circ d = 0.$$

∎

Let us finish this paragraph with a description of the de Rham cohomology : we consider the following sub-R-modules of Ω :

a)

$$
\begin{aligned}
Z^n(M; E) \ &= \ \text{Ker } d \\
&= \ \{\omega \in \Omega_n(M; E) \mid d\omega = 0\}
\end{aligned}
$$

b)

$$
\begin{aligned}
B^n(M; C) \ &= \ \text{Im } d \\
&= \ \{d\omega \mid \omega \in \Omega_{n-1}(M; E)\}
\end{aligned}
$$

By virtue of Proposition 6, $B^n(M; E)$ is a sub-R-module of $Z^n(M; E)$. We then put the following definition.

Definition 4. *The nth de Rham cohomology R-module with coefficients in E is the quotient*

$$H^n(M; E) = Z^n(M; E)/B^n(M; E).$$

4.3 Integral of differential forms

4.3.1 Integral of a n-form on an n-interval

In this paragraph we shall restrict ourselves to the study of differential forms with value in R. Furthermore we shall suppose that, in addition to the (K-W) axiom (2.1.3), R satisfies the axioms of order and integration (cf. 1.3.1 and 1.3.2).

Let P be an n-interval, in other words the product of n intervals

$$P = [a_1, b_1] \times [a_2, b_2] \times \ldots \times [a_n, b_n].$$

Let M be a microlinear object and

$$\tau : P \to M$$

be an arc of n-surface in M. Let ω be a differential n-form on M with value in R.

If $\underline{u} \in P$, we define the microcubes

$$\tau_{\underline{u}} : D^n \to M$$

by

$$\tau_{\underline{u}}(\underline{d}) = \tau(\underline{u} + \underline{d}).$$

We obtain a function

$$\tau^* \omega : P \to R$$

by putting

$$(\tau^* \omega)(\underline{u}) = \omega(\tau_{\underline{u}}).$$

Remember (cf. 1.3) that an interval $[a, b]$ does not determine its extremities. Thus we have to consider marked n-intervals, in other words, triples $(P, \underline{a}, \underline{b})$ where $P = \prod_{i=1}^{n} [a_i, b_i]$. Then an *arc of marked n-surface* is a triple $(\tau, \underline{a}, \underline{b})$ where $\tau : P \to M$ is an arc of n-surface in M and $P = \prod_{i=1}^{n} [a_i, b_i]$.

Definition 1. *Let $(\tau, \underline{a}, \underline{b})$ be an arc of marked n-surface. We call the integral of ω on $(\tau, \underline{a}, \underline{b})$, which we denote by*

$$\int_{(\tau, \underline{a}, \underline{b})} \omega,$$

the number

$$\int_{a_1}^{b_1} \ldots \int_{a_n}^{b_n} (\tau^*\omega)(\underline{u}) du_1 du_2 \ldots du_n.$$

We immediately have :

Proposition 1. $\int_{(\tau,\underline{a},\underline{b})} \omega$ *is an R-linear function of* ω. ∎

Let $M \xrightarrow{f} N$. If $\tau : P \to M$ is an arc of n-surface on M, we put $f_*\tau = f \circ \tau$ and it is an arc of n-surface on N. Thus we see that the set $S_n(M)$ of n-surfaces arcs on M furnishes a covariant functor in M. Remember also that $\Omega_n(M) = (\Omega_n(M; R))$ is a contravariant functor in M : if $\omega \in \Omega_n(N)$, $f^*\omega$ is defined by $f^*\omega(\gamma) = \omega(f \circ \gamma)$. Then we have easily :

Proposition 2. *For every* $f : M \to N$, *every* $\tau : P \to M$ *and every* $\omega \in \Omega_n(N)$, *we have*

$$\int_{(\tau,\underline{a},\underline{b})} f^*\omega = \int_{(f_*\tau,\underline{a},\underline{b})} \omega.$$

∎

We intend to show now that the integral notion just described does extend well the integral notion on marked n-microcubes that we saw in the previous section.

Let $(\gamma, \underline{e})$ be a marked n-microcube. We associate to it the arc of marked n-surface $\gamma^{\underline{e}}$ defined on $[0, 1]^n$ by

$$\gamma^{\underline{e}}(t_1, \ldots, t_n) = \gamma(e_1 t_1, \ldots, e_n t_n).$$

Proposition 3. *We have*

$$\int_{(\gamma,\underline{e})} \omega = \int_{(\gamma^{\underline{e}},\underline{0},\underline{1})} \omega.$$

Proof: The left hand side, which is defined in 4.2.2, is

$$e_1 \cdot e_2 \cdot \ldots e_n \cdot \omega(\gamma).$$

The right hand side is here

$$\int_0^1 \ldots \int_0^1 \omega((\gamma^{\underline{e}})_{\underline{u}}) du_1 \ldots du_n$$

where
$$(\gamma^{\underline{e}})_{\underline{u}}(\underline{d}) = \gamma(e_1(u_1 + d_1), \ldots, e_n(u_n + d_n)).$$

Let us consider the two functions given by
$$F(\underline{e}, \underline{e}') = \omega(\sigma_{\underline{e},\underline{e}'}),$$

where $\sigma_{\underline{e},\underline{e}'}$ is the n-microcube :
$$\sigma_{\underline{e},\underline{e}'}(\underline{d}) = \gamma(e_1'u_1 + e_1d_1, \ldots, e_n'u_n + e_nd_n);$$

and by
$$G(\underline{e}, \underline{e}') = e_1 \ldots e_n \omega(\gamma).$$

If $\underline{e}' = \underline{0}$ we get
$$\sigma_{\underline{e},\underline{0}}(\underline{d}) = \gamma(e_1d_1, \ldots, e_nd_n)$$

and, by the multihomogeneity of ω,
$$\begin{aligned} F(\underline{e}, \underline{0}) &= e_1 \ldots e_n \omega(\gamma) \\ &= G(\underline{e}, \underline{0}). \end{aligned}$$

If $\underline{e} = \underline{0}$ we obtain that $\sigma_{\underline{0},\underline{e}'}$ is constant, and thus $F(\underline{0}, \underline{e}') = 0$, and also $G(\underline{0}, \underline{e}')$.

By the general (K-W)-axiom, F and G coincide on $D(2)^n$, in particular on $(\underline{e}, \underline{e})$. So we have
$$\omega((\gamma^{\underline{e}})_{\underline{u}}) = e_1 \ldots e_n \omega(\gamma)$$

and the proposition is proved. ∎

The free R-module generated by the arcs of marked n-surfaces will be called the R-module of n-chains of M and denoted $\Gamma_n(M)$. The integral being defined on the arc of marked n-surface extends by linearity to the n-chains and thus we have a bilinear mapping
$$\int : \Gamma_n(M) \times \Omega_n(M) \to R.$$

4.3.2 Stokes formula

In section 4.2.1 we have defined the boundary operator for infinitesimal $(n+1)$-chains. We extend trivially this operation to $(n+1)$-chains :

$$\partial : \Gamma_{n+1}(M) \to \Gamma_n(M).$$

Actually it suffices to define the boundary of an arc of marked $(n+1)$-surface $(\tau, \underline{a}, \underline{b})$ by

$$\partial(\tau, \underline{a}, \underline{b}) = \sum_{i=1}^{n+1}(-1)^i F_{\underline{a}}^i(\tau, \underline{a}, \underline{b}) - \sum_{i=1}^{n+1}(-1)^i F_{\underline{b}}^i(\tau, \underline{a}, \underline{b})$$

where

$$F_{\underline{a}}^i(\tau, \underline{a}, \underline{b}) = (\tau_{\underline{a}}^i, \underline{a}_{(i)}, \underline{b}_{(i)})$$

with

$$\begin{aligned}
\tau_{\underline{a}}^i(u_1, \ldots, u_n) &= \tau(u_1, \ldots, u_{i-1}, a_i, u_i, \ldots, u_n) \\
\underline{a}_{(i)} &= (a_1, \ldots, \widehat{a}_i, \ldots, a_{n+1}) \\
\underline{b}_{(i)} &= (b_1, \ldots, \widehat{b}_i, \ldots, b_{n+1})
\end{aligned}$$

and where $F_{\underline{b}}^i(\tau, \underline{a}, \underline{b})$ is defined in an similar way.

Before we prove the Stokes formula, let us fix some notations allowing us to designate "slices" of marked arcs. Let $(\tau, \underline{a}, \underline{b})$ be an arc of marked $(n+1)$-surface, k an index $(1 \le k \le n+1)$, and x, y satisfy

$$a_k \le x \le y \le b_k.$$

By $(\tau, \underline{a}, \underline{b})_{x,y}^k$ we refer to the arc of marked $(n+1)$-surface where we replace, in the kth direction, a_k by x and b_k by y. Explicitly,

$$(\tau, \underline{a}, \underline{b})_{x,y}^k = (\tau_{x,y}^k, \underline{a}_x^k, \underline{b}_y^k)$$

where $\tau_{x,y}^k = \tau \mid_{P_{x,y}^k}$ with

$$\begin{aligned}
P_{x,y}^k &= [a_1, b_1] \times \ldots \times [a_{k-1}, b_{k-1}] \times [x, y] \times \ldots \times [a_{n+1}, b_{n+1}] \\
\underline{a}_x^k &= (a_1, \ldots, a_{k-1}, x, \ldots, a_{n+1}) \\
\underline{b}_y^k &= (b_1, \ldots, b_{k-1}, y, \ldots, b_{n+1}).
\end{aligned}$$

As we shall use the integration axiom in order to "sum" by infinitesimal slices, it is useful to observe that if

$$a_k \le x \le y \le z \le b_k$$

we have, for a differential $(n+1)$-form θ :

$$\int_{(\tau,\underline{a},\underline{b})^k_{x,z}} \theta = \int_{(\tau,\underline{a},\underline{b})^k_{x,y}} \theta + \int_{(\tau,\underline{a},\underline{b})^k_{y,z}} \theta$$

because for every function $F : [a_k, b_k] \to R$ we have

$$\int_x^z F = \int_x^y F + \int_y^z F$$

by virtue of (1.3, Prop. 10).

This additivity is also true on the bounds. More precisely, if ω is a n-form we have

$$\int_{\partial((\tau,\underline{a},\underline{b})^k_{x,z})} \omega = \int_{\partial((\tau,\underline{a},\underline{b})^k_{x,y})} \omega + \int_{\partial((\tau,\underline{a},\underline{b})^k_{y,z})} \omega.$$

In order to see this, it suffices to observe that for sides situated in directions other than the kth, it follows again from the additivity of the integral (1.3.3, Prop. 10), whereas for the sides situated in the direction k the additional terms of the righthand side that contain y appear with two opposite signs.

We can now prove the *Stokes formula* :

Proposition 4. *If* $c \in \Gamma_{n+1}(M)$ *and* $\omega \in \Omega_n(M)$, *we have* :

$$\int_{\partial c} \omega = \int_c d\omega.$$

Proof: We can limit ourselves to the instance where c is an arc of marked $(n+1)$-surface $(\tau, \underline{a}, \underline{b})$.

a) If $b_i - a_i \in D$ for each i $(1 \le i \le n+1)$, the proposition is true by definition of $d\omega$. In fact, we then associate to $(\tau, \underline{a}, \underline{b})$ the $(n+1)$-microcube $(\gamma, \underline{e})$ where $\gamma(\underline{d}) = \tau(\underline{a} + \underline{d})$ and $\underline{e} = \underline{b} - \underline{a}$. Taking into account Proposition 3, we can then use Proposition 3 of 4.2.3.

b) Suppose that the proposition is true when $b_i - a_i \in D$ for $k \le i \le n+1$ and let us show that it still is true when we have only $b_i - a_i \in D$ for $k+1 \le i \le n+1$.

Let $(\tau, \underline{a}, \underline{b})$ be of this type. Define two functions

$$g \;:\; [a_k, b_k] \to R$$
$$h \;:\; [a_k, b_k] \to R$$

by :

$$h(x) \;=\; \int_{(\tau,\underline{a},\underline{b})^k_{a_k,x}} d\omega$$

$$g(x) \;=\; \int_{\partial((\tau,\underline{a},\underline{b})^k_{a_k,x})} \omega.$$

It suffices to show that $h(b_k) = g(b_k)$. But this will follow from the uniqueness clause of the integral on an interval (cfr. 1.3, Prop. 7) if we prove that $h(a_k) = g(a_k)$ and $h'(x) = g'(x)$.

We first have $h(a_k) = g(a_k) = 0$ because an integral of the form $\int_a^a F$ is always zero (cf. 1.3.3, Prop. 11).

Compute $h'(x)$ and $g'(x)$. We have for $d \in D$,

$$h(x + d) - h(x) \;=\; \int_{(\tau,\underline{a},\underline{b})^k_{a_k,x+d}} d\omega - \int_{(\tau,\underline{a},\underline{b})^k_{a_k,x}} d\omega$$

$$=\; \int_{(\tau,\underline{a},\underline{b})^k_{x,x+d}} d\omega.$$

Indeed $(\tau, \underline{a}, \underline{b})^k_{x,x+d}$ is not only infinitesimal in the directions $k+1$, $\dots, n+1$ but also in the direction k. By inductive hypothesis the Stokes formula applies already to this kind of an arc of marked $(n+1)$-surface and we have

$$h(x + d) - h(x) \;=\; \int_{\partial((\tau,\underline{a},\underline{b})^k_{x,x+d})} \omega$$

$$=\; g(x + d) - g(x).$$

Thus we have for every d in D, $d \cdot h'(x) = d \cdot g'(x)$ and thus $h'(x) = g'(x)$. So we do get

$$h(b_k) = g(b_k).$$

We can thus step-by-step decrease the number of directions in which the $n + 1$-intervals have infinitesimal sides and after $(n + 1)$ steps the Stokes formula is established for all intervals. ∎

4.4 The exterior algebra of differential forms

4.4.1 Multilinear forms

Let $\lambda : M^{D^p} \to R$ be a p-homogeneous mapping. We say that it is a p-linear form on M. We denote by $L_p(M)$ the set of p-linear forms on M. It is obviously an R-module and also an R^M-module (cf. 4.1.1, prop. 3).

Definition 1. Let $\lambda \in L_p(M)$ and $\mu \in L_q(M)$. We define the tensor product of λ and μ, which we denote by $\lambda \otimes \mu$, to be the mapping

$$(\lambda \otimes \mu) : M^{(D^{p+q})} \to R, \ (\lambda \otimes \mu)(\gamma) = \lambda(\gamma(-,\underline{0})).\mu(\gamma(\underline{0},-)).$$

Trivially we verify the following properties :

Proposition 1. Let $\lambda \in L_p(M)$, $\mu \in L_q(M)$ and $\nu \in L_r(M)$. We have

1) $\lambda \otimes \mu \in L_{p+q}(M)$;

2) the tensor product

$$\otimes \ : \ L_p(M) \times L_q(M) \to L_{p+q}(M)$$

is bilinear ;

3) $(\lambda \otimes \mu) \otimes \nu = \lambda \otimes (\mu \otimes \nu).$

∎

We evidently have
$$\Omega_p(M) \subseteq L_p(M).$$
In fact this inclusion has a retraction . Let $\lambda \in L_p(M)$.

Definition 2. The anti-symmetrization of λ is the mapping $A(\lambda) = M^{D^p} \to R$ defined by :

$$A(\lambda)(\gamma) = \sum_\sigma \epsilon_\sigma \lambda(\gamma \circ \sigma)$$

where $\sigma \in S_p$ (the set of permutation from $\{1,\ldots,p\}$).

An easy computation gives the following properties :

Proposition 2.

1) *A is a linear mapping from $L_p(M)$ to $\Omega_p(M)$.*

2) *If $\omega \in \Omega_p(M)$, $A(\omega) = p!\omega$.*

3) *If $\lambda \in L_p(M)$ and $\mu \in L_q(M)$,*

we have

$$A(\lambda \otimes \mu) = (-1)^{p.q} A(\mu \otimes \lambda).$$

∎

Let us note the functorial properties . Let $f : M \to N$. We define $f^* : L_p(N) \to L_p(M)$ by putting

$$(f^*\lambda)(\gamma) = \lambda(f \circ \gamma).$$

Proposition 3.

(1) *L_p is thus a contravariant functor.*

(2) *For every f, f^* is linear.*

(3) *$f^*(A(\lambda)) = A(f^*\lambda)$*

(4) *$f^*(\lambda \otimes \mu) = f^*\lambda \otimes f^*\mu$.*

∎

4.4.2   The exterior product

The tensor product of two differential forms of degree p and q is also a $(p+q)$-linear form but it is no longer alternated. In order to get a differential form, we apply the anti-symmetrization operator.

Definition 3.   Let $\omega_1 \in \Omega_p(M)$ and $\omega_2 \in \Omega_q(M)$. The exterior product of ω_1, ω_2 is the differential form of degree $p + q$:

$$\omega_1 \wedge \omega_2 = A(\omega_1 \otimes \omega_2) \cdot \frac{1}{p!\,q!}.$$

The algebraic properties of the exterior product can then be etablished by reducing ourselves to an easy computation (sometimes tedious) about permutations and their signatures.

Proposition 4. *The exterior product is a bilinear mapping*

$$\wedge \; : \; \Omega_p(M) \times \Omega_q(M) \to \Omega_{p+q}(M).$$

The exterior product is graded-commutative, i.e. if

$$\omega_1 \in \Omega_p(M) \;\; \text{and} \;\; \omega_2 \in \Omega_q(M),$$

$$\omega_1 \wedge \omega_2 = (-1)^{p \cdot q} \omega_2 \wedge \omega_1.$$

The exterior product is associative. If $f : M \to N$, we have

$$f^*(\omega_1 \wedge \omega_2) = f^*\omega_1 \wedge f^*\omega_2.$$

∎

Let us observe that the differential forms of degree 0 are simply the functions from M to $R : \Omega_0(M) = R^M$. The exterior product of a form of degree 0 with a form of degree p is simply the ordinary product

$$\Omega_0(M) \times \Omega_p(M) \to \Omega_p(M)$$

given by the R^M-module structure on $\Omega_p(M)$.
We write

$$\Omega(M)$$

for the direct sum of the $\Omega_p(M)$ for all p in $\mathbb{N}$. It is called the algebra of differential forms. Its structure is very simple.

Proposition 5. *The algebra $\Omega(M)$ is an associative graded-commutative graded algebra.* ∎

We say that a differential form of degree p is decomposable if it can be written as

$$\omega = \omega_1 \wedge \omega_2 \wedge \ldots \wedge \omega_p$$

where the ω_i are degree one forms. We can show that the R-module generated by the decomposable differential forms is the R-module of classical differential forms (cfr. Section 4.1.2).

Let us point out the interaction with the exterior differential. By using Propositions 3 and 4 of 4.2.3 and the definition of exterior product, a rather courageous reader could verify the following proposition :

Proposition 6. *If $\omega_1 \in \Omega_p(M)$ and $\omega_2 \in \Omega_q(M)$, we have :*

$$d(\omega_1 \wedge \omega_2) = d\omega_1 \wedge \omega_2 + (-1)^p \omega_1 \wedge d\omega_2.$$

■

The less courageous reader can find explicit proof in C. Minguez Herrero [47].

4.4.3   Interior products. Lie derivatives

Let $X : D \times M \to M$ be a vector field on M and let $\omega : M^{D^p} \to R$ be a p-degree differential form on M. We are going to define a $p - 1$ degree differential form associated to the couple (X, ω).

First the X-field is made to act on the $(p-1)$-microcubes to transform them into p-microcubes in the following way. If $\gamma : D^{p-1} \to M$, we define

$$X * \gamma : D^p \to M$$

by $(X * \gamma)(d_1, \ldots, d_p) = X_{d_1}(\gamma(d_2, \ldots, d_p))$.

Then we make the definition :

Definition 4. *The interior product of a differential p-form ω with a vector field X, denoted $i_X \omega$, is the map*

$$i_X \omega : M^{D^{p-1}} \to R$$

given by $(i_X \omega)(\gamma) = \omega(X * \gamma)$.

Proposition 7. *For every X and every ω, $i_X \omega$ is a differential $(p-1)$-form. The mapping*

$$\mathcal{X}(M) \times \Omega_p(M) \to \Omega_{p-1}(M)$$

$$(X, \omega) \rightsquigarrow i_X \omega$$

is bilinear. And we have for $\omega_1 \in \Omega_p(M)$ and $\omega_2 \in \Omega_q(M)$

$$i_X(\omega_1 \wedge \omega_2) = i_X \omega_1 \wedge \omega_2 + (-1)^p \omega_1 \wedge i_X \omega_2.$$

Proof: As

$$(X * \gamma)(d_1, d_2, \ldots, d_p) = X_{d_1}(\gamma(d_2, \ldots, d_p))$$

and as ω is p-homogeneous and alternated, $i_X\omega$ is also $(p-1)$-homogeneous and alternated, these properties depending only on the variables $(d_2, \ldots, d_p)$. The homogeneity of ω at the variable d_1 expresses the linearity of $i_X\omega$ in X. The linearity in ω is immediate.

It remains to prove the last formula, expressing that i_X is an anti-differentiation. In order to simplify notation, we shall denote the microcubes by specifying the variables. If σ is a permutation of the numbers $(1, \ldots, p+q)$ we shall denote for instance by $\sigma(d_i, d_j, \ldots)$ the sequence $(d_{\sigma(i)}, d_{\sigma(j)}, \ldots)$. Then we have

$$i_X(\omega_1 \wedge \omega_2)(\gamma(d_2, d_3, \ldots, d_{p+q}))$$
$$= (\omega_1 \wedge \omega_2)(X_{d_1}(\gamma(d_2, d_3, \ldots, d_{p+q})))$$
$$= \frac{1}{p!}\frac{1}{q!} \sum_{\sigma \in S_{p+q}} \epsilon_\sigma(\omega_1 \otimes \omega_2)((X * \gamma) \circ \sigma)$$
$$= \frac{1}{p!}\frac{1}{q!} \sum_{\sigma \in S_{p+q}} \epsilon_\sigma \omega_1((X * \gamma)(\sigma(d_1, \ldots, d_p, 0, \ldots, 0)))$$
$$. \; \omega_2((X * \gamma)(\sigma(0, 0, \ldots 0, d_{p+1}, \ldots, d_{p+q})))$$

We look for index 1 : let $k = \sigma^{-1}(1)$. Put

$$S' = \{\sigma \in S_{p+q} \mid k \leq p\} \text{ and } S'' = \{\sigma \in S_{p+q} \mid k > p\}$$

we have $S_{p+q} = S' \cup S''$ and we can split up the previous sum into two sums over the σ of S' and of S'' respectively. The first of these two sums is

$$\frac{1}{p!}\frac{1}{q!} \sum_{\sigma \in S'} \epsilon_\sigma \cdot \omega_1((X * \gamma)(\sigma(d_1, \ldots, d_p, 0, \ldots, 0)))$$
$$. \; \omega_2((X * \gamma)(0, 0, \ldots, 0, d_{p+1}, \ldots, d_{p+q})))$$

Let τ_σ be the transposition exchanging 1 and $k = \sigma^{-1}(1)$. We get ;

$$\frac{1}{p!}\frac{1}{q!} \sum_{\sigma \in S'} \epsilon_{\tau_\sigma . \sigma} \, \omega_1 \left((X * \gamma)(d_1, \tau_\sigma \sigma(d_2, \ldots, d_p, 0, 0, \ldots, 0)) \right)$$

$$\cdot \, \omega_2 \left((X * \gamma)(0, \tau_\sigma \sigma(0, \dots, d_{p+1}, \dots, d_{p+q})) \right)$$

$$= \frac{1}{p!} \frac{1}{q!} \sum_{\sigma \in S'} \epsilon_{\tau_\sigma \cdot \sigma} (i_X \omega_1) \left(\gamma(\tau_\sigma \sigma(d_2, \dots, d_p, 0, 0, \dots, 0)) \right)$$

$$\cdot \, \omega_2 \left(\gamma(\tau_\sigma \sigma(0, \dots, 0, d_{p+1}, \dots, d_{p+q})) \right).$$

We identify the $\rho \in S_{p+q-1}$ with the permutations of $(1, 2, \dots, p+q)$ fixing 1 . For fixed ρ , there exist p permutations σ of S' such that $\rho = \tau_\sigma \sigma$ and thus we get

$$= p \frac{1}{p!} \frac{1}{q!} \sum_{\rho \in S_{p+q-1}} \epsilon_\rho (i_X \omega_1)(\gamma \circ \rho) \cdot \omega_2(\gamma \circ \rho)$$

$$= (i_X \omega_1 \wedge \omega_2)(\gamma).$$

Similarly we show that the second sum (over the σ of S'') is equivalent to

$$(-1)^p (\omega_1 \wedge i_X \omega_2)(\gamma).$$

∎

If u is a bijective map from M to M, we define $u^*(\omega)$ by $u^*(\omega)(\gamma) = \omega(u \circ \gamma)$ and $u^*(\omega)$ is a p-form as ω is. If X is a vector field on M, then X_d is a bijective map from M to M. We have thus the following definition.

Definition 5. *Let X be a vector field on M and $\omega \in \Omega_p(M)$. We define the* Lie derivative *of ω in the direction of X, denoted by $L_X \omega$, by*

$$(X_d)^*(\omega)(\gamma) - \omega(\gamma) = d \cdot (L_X \omega)(\gamma)$$

for all $d \in D$ and all microcubes γ.

We give here a proposition characterizing L_X in terms of i_X and of the exterior differential.

Proposition 8. *Let X be a vector field on M and $\omega \in \Omega_p(M)$. We have*

$$L_X \omega = (i_X d + d i_X)(\omega).$$

Proof: We compute $i_X d\omega(\gamma)$ for $\gamma : D^p \longrightarrow M$. By the explicit definition of the exterior differential (4.2.2, proposition 4) we have

$$(i_X d\omega)(\gamma) = d\omega(X * \gamma)$$

$$= \sum_{i=1}^{n+1}(-1)^{i+1}DF^i(0)$$

where $F^i(e) = \omega((X * \gamma)^i(e))$ with $(X * \gamma)^i(e) = (X * \gamma)(d_1,\ldots,d_p)$.
We isolate the first term $DF^1(0)$. For $e \in D$,

$$
\begin{aligned}
e \cdot DF^1(0) &= F^1(e) - F^1(0) \\
&= \omega((X * \gamma)^1(e)) - \omega((X * \gamma)^1(0)) \\
&= \omega(X_e(\gamma(d_1,\ldots,d_p))) - \omega(\gamma(d_1,\ldots,d_p)) \\
&= (X_e)^*(\omega)(\gamma) - \omega(\gamma) \\
&= e \cdot (L_X\omega)(\gamma)
\end{aligned}
$$

(we denote the variables by $d_1,\ldots,d_p$) .We have thus

$$DF^1(0) = (L_X\omega)(\gamma).$$

The other terms are $\sum_{i=2}^{p+1}(-1)^{i+1}DF^i(0)$.
    We compute now $(di_X\omega)(\gamma)$. By the same explicit expression of the exterior differential we have

$$(di_X\omega)(\gamma) = \sum_{i+1}^{p}DG^i(0)$$

where

$$
\begin{aligned}
G^i(e) &= (i_X\omega)(\gamma^i(e)) \\
&= \omega(X * \gamma^i(e)).
\end{aligned}
$$

Then, comparing F^i and G^i, we get $DG^i(0) = DF^{i+1}(0)$ and the proof is complete. ∎

    The properties of d and i_X give immediately the following proposition.

Proposition 9.

(1) $L_X\omega$ is a differential p-form .

(2) The mapping

$$L : \mathcal{X}(M) \times \Omega_p(M) \to \Omega_p(M)$$

$$(X, \omega) \rightsquigarrow L_X \omega$$

is bilinear.

(3) If $\omega_1 \in \Omega_p(M)$ and $\omega_2 \in \Omega_q(M)$, then

$$L_X(\omega_1 \wedge \omega_2) = L_X \omega_1 \wedge \omega_2 + \omega_1 \wedge L_X \omega_2.$$

∎

Instead of making the vector fields operate (by interior products) on forms, we can also make them operate on higher dimensional distributions.

Let us consider the mapping

$$M^{D^q} \to M$$

$$\gamma \rightsquigarrow \gamma(\underline{0}).$$

A section of that mapping is called a q-dimensional distribution (cf. end of 3.4). We can also consider it as a mapping

$$\mathcal{D}_q : D^q \times M \to M$$

such that $\mathcal{D}_q(\underline{0}, x) = x$. Then let $\omega \in \Omega_p(M)$ with $p \geq q$.

Definition 6. Let $\omega \in \Omega_p(M)$ and let $\mathcal{D}_q$ be a q-dimensional distribution with $q \leq p$. The interior product $i_{\mathcal{D}_q} \omega$ is the map

$$M^{D^{p-q}} \to R$$

given by

$$(i_{\mathcal{D}_q} \omega)(\gamma) = \omega(\mathcal{D}_q * \gamma)$$

where

$$(\mathcal{D}_q * \gamma)(d_1, \ldots, d_p) = \mathcal{D}_q(d_1, \ldots, d_q)(\gamma(d_{q+1}, \ldots, d_p)).$$

It is clear that $i_{\mathcal{D}_q} \omega$ is a $p - q$ degree differential form.

4.5 De Rham's Theorem

We introduce here a notation frequently used in the sequel. A character x decorated with a lower tilde $\underset{\sim}{x}$ is considered as a variable ranging over some set fixed by the context. For example, given $f : D \times R \to R$, $f(\underset{\sim}{d}, t)$ is the function from D to R sending d to $f(d, t)$ for fixed t.

4.5.1 Integration as a cochain morphism

Let M be a microlinear object. We have the differential complex of differential forms : $(\Omega_p(M), d)$.

A p-<u>cube</u> τ in M is a morphism $\tau : [0, 1]^p \to M$. The free R-module generated by the p-cubes of M is denoted by $\Gamma_p(M)$. The integral can be extended by linearity:

$$\int : \Gamma_p(M) \times \Omega_p(M) \to R$$

Definition 1. *The elements c and c' of $\Gamma_p(M)$ are called equivalent if we have*

$$\int_c \omega = \int_{c'} \omega$$

for each ω in $\Omega_p(M)$.

We denote by $S_p(M)$ the quotient module of $\Gamma_p(M)$ by this equivalence relation and by $S^p(M)$ the dual module $\mathcal{L}_R(S_p(M), R)$. The elements of $S^p(M)$ are called *cubical singular p-cochains* on M.

We have the boundary operator

$$\partial \; : \Gamma_{p+1}(M) \to \Gamma_p(M),$$

which gives us

$$\partial \; : \; S_{p+1} \to S_p(M),$$

and we obtain

$$\delta \; : \; S^p(M) \to S^{p+1}(M)$$

by $\delta(\phi)(c) = \phi(\partial c))$.

Integration gives us a morphism

$$\int : \; \Omega_p(M) \to S^p(M).$$

The Stokes formula (4.3 proposition 4) implies the following proposition.

Proposition 1. *For every p the diagram*

$$
\begin{CD}
\Omega_p(M) @>\int>> S^p(M) \\
@VdVV @VV\delta V \\
\Omega_{p+1}(M) @>\int>> S^{p+1}(M)
\end{CD}
$$

commutes. Thus $\int$ is a morphism of cochain complexes.

4.5.2   The differentiation morphism D.

We now define a morphism

$$ D \; : \; S^p(M) \to \Omega_p(M). $$

Let ϕ belong to $S^p(M)$. We have

$$ \Psi \; : \; M^{D^p} \times D^p \to R $$

$$ \Psi(\gamma, \underline{e}) \; = \; \phi(\gamma^{\underline{e}}) $$

where $\gamma^{\underline{e}} \; : \; [0,1]^p \to M$ is given by $\gamma^{\underline{e}}(\underline{t}) = \gamma(\underline{t} \cdot \underline{e})$ and $\underline{t} \cdot \underline{e} = (t_1 \cdot e_1, \cdots, t_p \cdot e_p)$.

Proposition 2. *There exists a unique differential p-form $D\phi$ which satisfies*

$$ e_1 \cdots \cdots e_p \cdot (D\phi)(\gamma) = \phi(\gamma^{\underline{e}}). $$

Proof: The result will follows by proposition 2 of 4.2.2 if only we can prove that Ψ satisfies the hypotheses of that proposition.

(a) Homogeneity in the variable d_i. We have to show

$$ \Psi(\alpha \underset{i}{\cdot} \gamma, \underline{e}) = \alpha \cdot \Psi(\gamma, \underline{e}), $$

i.e. $(\alpha \underset{i}{\cdot} \gamma)^{\underline{e}} \sim \alpha \cdot \gamma^{\underline{e}}$ (where $\sim$ denotes the equivalence of chains). Let ω be a p-form. It is enough to show that

$$ \int_{(\alpha \underset{i}{\cdot} \gamma)^{\underline{e}}} \omega = \alpha \cdot \int_{\gamma^{\underline{e}}} \omega $$

but this follows directly from proposition 3 of 4.3.

(b) Alternating condition. Let s be a permutation of $1, \ldots, p$. We have $(\sigma \circ s)^e \sim \epsilon_s \cdot \sigma^e$. In fact for any ω

$$\int_{(\sigma \circ s)^e} \omega = e_1 \cdots e_p \omega(\sigma \circ s) = \epsilon_s \, e_1 \cdots e_p \omega(\sigma) = \epsilon_s \int_{\sigma^e} \omega.$$

(c) Homogeneity in the e_i is evident. ∎

We thus have a morphism

$$D \; : \; S^p(M) \to \Omega_p(M).$$

Proposition 3. *For every p the diagram*

$$
\begin{array}{ccc}
S^p(M) & \xrightarrow{\;\;D\;\;} & \Omega_p(M) \\
{\scriptstyle \delta}\downarrow & & \downarrow{\scriptstyle d} \\
S^{p+1}(M) & \xrightarrow{\;\;D\;\;} & \Omega_{p+1}(M)
\end{array}
$$

commutes.

Proof: (We use here the notation with variables as explained at the beginning of this section). Let $\underline{e}$ be in D^{p+1} and τ in $M^{D^{p+1}}$. We have

$$e_1 \cdots e_{p+1} \cdot d(D(\phi))(\tau(d_1, \cdots, d_{p+1}))$$

$$= \sum_{i=1}^{p+1} [\phi(\tau(t_1 e_1, \cdots, t_{i-1} e_{i-1}, e_i, t_i e_{i+1}, \cdots, t_p e_{p+1}))$$

$$- \phi(\tau(t_1 e_1, \cdots, t_{i-1} e_{i-1}, \; 0 \; , t_i e_{i+1}, \cdots, t_p e_{p+1}))]$$

On the other hand

$$e_1 \cdots e_{p+1} \cdot D(\delta \phi)(\tau(d_1, \cdots, d_{p+1}))$$

$$= (\delta \phi)(\tau(t_1 \cdot e_1, \cdots, t_{p+1}))$$

$$= \phi(\partial(\tau(t_1 e_1, \cdots, t_{p+1} e_{p+1})))$$

$$= \phi(\sum_{i=1}^{p+1} [\tau(t_1 e_1, \cdots, t_{i-1} e_{i-1}, \; 0 \; t_{i+1} e_{i+1}, \cdots, t_{p+1} e_{p+1}$$

$$- \tau(t_1 e_1, \cdots, t_{i-1} e_{i-1}, \; e_i \;, t_{i+1} e_{i+1}, \cdots, t_{p+1} e_{p+1})]),$$

and therefore $d \circ D = D \circ \delta$. ∎

Now we verify that D and $\int$ are inverse isomorphisms.

Proposition 4.

$$D \circ \int = id.$$

Proof: Let ω be a differential p-form and $\underline{e}$ an element of D^p. We have

$$e_1 \cdots e_p \cdot D\left(\int \omega\right)(\sigma) = \int \omega(\sigma^e) = \int_{\sigma e} \omega = e_1 \cdots e_p \cdot \omega(\sigma).$$

(by proposition 3 in 4.3). ∎

We want now to show that, if ϕ belongs to $S^p(M)$, σ to $S_p(M)$, and $\underline{e}$ to D^p, the following equality holds :

$$e_1 \cdots e_p \cdot \int (D(\phi))(\sigma) = e_1 \cdots e_p \cdot \phi(\sigma).$$

Let us consider the two functions F and G from $[0,1]^p$ to R defined by

$$F(\alpha_1, \ldots, \alpha_p) = e_1 \cdots e_p \cdot \int D(\phi)(\sigma(\alpha_1 \cdot \underline{t}_1, \ldots, \alpha_p \cdot \underline{t}_p)),$$
$$G(\alpha_1, \ldots, \alpha_p) = e_1 \cdots e_p \cdot \phi(\sigma(\alpha_1 \cdot \underline{t}_1, \ldots, \alpha_p \cdot \underline{t}_p)).$$

We have

$$F(\alpha_1, \ldots, \alpha_p)$$
$$= e_1 \cdots e_p \cdot \int_0^1 \cdots \int_0^1 D(\phi)(\sigma(\alpha_1(u_1 + \underline{d}_1),$$
$$\ldots, \alpha_p(u_p + \underline{d}_p)))du_1 \cdots du_p$$
$$= e_1 \cdots e_p \cdot \int_0^{\alpha_1} \cdots \int_0^{\alpha_p} D(\phi)(\sigma(v_1 + \underline{d}_1, \ldots, v_p + \underline{d}_p))dv_1 \cdots dv_p$$
$$= \int_0^{\alpha_1} \cdots \int_0^{\alpha_p} \phi(\sigma(v_1 + e_1 \cdot \underline{t}_1, \ldots, v_p + e_p \cdot \underline{t}_p))dv_1 \cdots dv_p.$$

Lemma 5. *If one of the α_i is zero, then $F(\alpha_1, \ldots, \alpha_p) = 0$. We have*

$$\frac{\partial^p F}{\partial x_1 \cdots \partial x_p}(\alpha_1, \ldots, \alpha_p) = \phi(\sigma(\alpha_1 + e_1 \cdot \underline{t}_1, \ldots, \alpha_p + e_p \cdot \underline{t}_p)).$$

Proof: This results from the above expression for $F(\alpha_1, \ldots, \alpha_p)$. ∎

Lemma 6 (Finite additivity). *For α in $[0,1]$, we have*

$$\sigma((\alpha+d)\cdot t_1,t_2,\ldots,t_p) - \sigma(\alpha t_1,t_2,\ldots,t_p) \sim \sigma(\alpha+d\cdot t_1,t_2,\ldots,t_p),$$

where $\sim$ denotes the equivalence relation for p-chains.

Proof: . Let ω be a p-form. We show that the integral of ω on σ_1 given by $\sigma_1(t_1,\ldots,t_p) = \sigma((\alpha+d)\cdot t_1,t_2,\ldots,t_p) - \sigma(\alpha t_1,t_2,\ldots,t_p)$ and on $\sigma_2 = \sigma(\alpha+d\cdot t_1,t_2,\ldots,t_p)$ are equal.

$$
\begin{aligned}
\int_{\sigma_1}\omega &= \int_0^1\cdots\int_0^1 [\omega(\sigma((\alpha+d)\cdot(s_1+d_1),s_2+d_2,\ldots,s_p+d_p)) \\
&\quad -\omega(\sigma(\alpha\cdot(s_1+d_1),s_2+d_2,\ldots,s_p+d_p))]\cdot ds_1\cdots ds_p \\
&= \int_0^1\cdots\int_0^1 [\omega(\sigma((\alpha+d)\cdot s_1+d_1),s_2+d_2,\ldots, \\
&\quad s_p+d_p))(\alpha+d)\cdot ds_1 ds_2\cdots ds_p \\
&\quad -\int_0^1\cdots\int_0^1 \omega(\sigma(\alpha\cdot s_1+d_1),s_2+d_2,\ldots,s_p+d_p))\alpha\cdot ds_1\,ds_2\cdots ds_p \\
&= \int_\alpha^{\alpha+d}\int_0^1\cdots\int_0^1 [\omega(\sigma(v_1+d_1,s_2+d_2,\ldots,s_p+d_p))]dv_1 ds_2\cdots ds_p \\
&= d\cdot\int_0^1\cdots\int_0^1 [\omega(\sigma(\alpha+d_1,s_2+d_2,\ldots,s_p+d_p))]ds_2\cdots ds_p.
\end{aligned}
$$

On the other hand,

$$
\begin{aligned}
\int_{\sigma_1}\omega &= \int_0^1\cdots\int_0^1 [\omega(\sigma(\alpha+d(s_1+d_1),s_2+d_2, \\
&\quad \ldots,s_p+d_p))]ds_1\,ds_2\cdots ds_p \\
&= \int_0^1\cdots\int_0^1 [\omega(\sigma(\alpha+d\cdot s_1+d_1,s_2+d_2, \\
&\quad \ldots,s_p+d_p))]\alpha\cdot ds_1\,ds_2\cdots ds_p \\
&= \int_0^d\int_0^1\cdots\int_0^1 (\omega(\sigma(\alpha+v_1+d_1,s_2+d_2,\ldots,s_p+d_p)))dv_1 ds_2\cdots ds_p \\
&= d\cdot\int_0^1\cdots\int_0^1 [\omega(\sigma(\alpha+d_1,s_2+d_2,\ldots,s_p+d_p))]ds_2\cdots ds_p.
\end{aligned}
$$

And the lemma is proved. ∎

Let us come back to $G(\alpha_1,\ldots,\alpha_p) = 0$

Lemma 7. (1) *If one of the α_i is zero, then $G(\alpha_1, \ldots, \alpha_p) = 0$, and*

$$(2)\frac{\partial^p G}{\partial x_1 \cdots \partial x_p}(\alpha_1, \ldots, \alpha_p) = \phi(\sigma(\alpha_1 + e_1 \cdot \underline{t}_1, \ldots, \alpha_p + e_p \cdot \underline{t}_p)).$$

Proof: . (1) This results clearly from the equivalence $\sigma(\alpha_1 \cdot \underline{t}_1, \ldots, \alpha_p \cdot \underline{t}_p) \sim 0$ valid if one of the α_i is null.

(2) results from Lemma 6 :

$$G(\alpha_1 + d_1, \alpha_2 + d_2, \ldots, \alpha_p + d_p) - G(\alpha_1, \ldots, \alpha_p)$$
$$= e_1 \cdots e_p \phi(\sigma(\alpha_1 + d_1 \cdot \underline{t}_1, \alpha_2 + d_2 \cdot \underline{t}_2, \ldots \alpha_p + d_p \cdot \underline{t}_p))$$
$$= d_1 \cdots d_p \phi(\sigma(\alpha_1 + e_1 \cdot \underline{t}_1, \alpha_2 + e_2 \cdot \underline{t}_2, \ldots \alpha_p + e_p \cdot \underline{t}_p)),$$

because the two p-cubes $e_1 \cdots e_p \sigma(\alpha_1 + d_1 \cdot \underline{t}_1, \alpha_2 + d_2 \cdot \underline{t}_2, \ldots \alpha_p + d_p \cdot \underline{t}_p)$ and $d_1 \cdots d_p \sigma(\alpha_1 + e_1 \cdot \underline{t}_1, \alpha_2 + e_2 \cdot \underline{t}_2, \ldots, \alpha_p + e_p \cdot \underline{t}_p)$ are equivalent. ■

Theorem 8. *D and $\int$ induce inverse isomorphisms between (Ω^p, d) and (S^p, δ).*

Proof: . We deduce from lemmas 5 and 7 that $F(\alpha_1, \ldots, \alpha_p) = G(\alpha_1, \ldots, \alpha_p)$. In particular for $(\alpha_1, \ldots, \alpha_p) = (1, 1, \ldots, 1)$, this gives

$$\int D(\phi)(\sigma) = \phi(\sigma).$$

This combined with proposition 4 gives the result. ■

We have thus a De Rham's theorem at the level of chain complexes.

4.5.3   Complements

By definition two p-chains are equivalent if they agree as currents. More precisely a p-current is a linear form on $\Omega^p(M)$ and a p-chain c acts as a p-current by $c(\omega) = \int_c \omega$. We could hope for an equivalence relation that does not use differential forms and integration, which is what we describe here.

Definition 2. *The degenerate p-chains are the chains generated by the p-chains of one of the following three kinds :*

1. The flat p-cubes $\tau : [0, 1]^p \to M$ that factor through a projection.

2. (Antisymmetry condition) The p-chains of the form $\sigma \circ s - \epsilon_s \cdot \sigma$, where s is a permutation of the set $1, 2, \ldots, p$ and ϵ_s is the signature of s.

3. (Finite additivity condition) For any p-cube σ and for any pair of elements (α, β) in $[0, 1]$ such that $\alpha + \beta$ belongs to $[0, 1]$, the p-chain

$$\sigma((\alpha+\beta) \cdot \underline{t}_1, \underline{t}_2, \ldots, \underline{t}_p) - \sigma(\alpha \cdot \underline{t}_1, \underline{t}_2, \ldots, \underline{t}_p) - \sigma(\alpha + \beta \cdot \underline{t}_1, \underline{t}_2, \ldots, \underline{t}_p).$$

We first observe stability of the degenerate chains for the boundary operator.

Proposition 9. The degenerate p-chains generate a sub-complex (D_p, ∂) of (Γ_p, ∂).

In other words, the boundary of a degenerate $(p + 1)$-chain is a degenerate p-chain.

Proof:
(a) If $\tau(t_1, \ldots, t_{p+1})$ is flat, we consider

$$\partial \tau = \sum_{i=1}^{p+1} (-1)^i \tau(\underline{s}_1, \ldots, \underline{s}_{i-1}, 0, \underline{s}_i, \ldots, \underline{s}_p)$$

$$- \sum_{i=1}^{p+1} (-1)^i \tau(\underline{s}_1, \ldots, \underline{s}_{i-1}, 1, \underline{s}_i, \ldots, \underline{s}_p).$$

If $\tau = \tau' \circ \pi_j$, the terms with $i \neq j$ are flat and the two terms with $i = j$ cancel.

(b) Consider the antisymmetry condition. Take for s the transposition of the two first variables:

$$\tau(t_2, t_1, t_3, \ldots, t_{p+1}) + \tau(t_1, t_2, t_3, \ldots, t_{p+1}).$$

The boundary is the sum

$$\tau \quad (\quad s, 0, s_2, \ldots, s_p) - \tau(s, 1, s_2, \ldots, s_p)$$
$$- \quad \tau(0, s, s_2, \ldots, s_p) + \tau(1, s, s_2, \ldots, s_p)$$

$$+\quad \tau(0,s,s_2,\ldots,s_p) - \tau(1,s,s_2,\ldots,s_p)$$
$$-\quad \tau(s,0,s_2,\ldots,s_p) + \tau(s,1,s_2,\ldots,s_p)$$
$$+\quad \sum_{i>2}(-1)^i\tau(s_2,s_1,\ldots,0,\ldots,s_p) - \sum_{i>2}(-1)^i\tau(s_2,s_1,\ldots,1,\ldots,s_p)$$
$$+\quad \sum_{i>2}(-1)^i\tau(s_1,s_2,\ldots,0,\ldots,s_p) - \sum_{i>2}(-1)^i\tau(s_1,s_2,\ldots,1,\ldots,s_p).$$

The first terms are killed two by two and the others are degenerate p-chains by antisymmetry.

(c) Let us consider now a $(p+1)$-degenerate chain of the form

$$\sigma((\alpha+\beta)\cdot t_1,t_2,\ldots,t_{p+1}) - \sigma(\alpha\cdot t_1,t_2,\ldots,t_{p+1}) - \sigma(\alpha+\beta\cdot t_1,t_2,\ldots,t_{p+1}).$$

The boundary is composed of degenerate p-chains of the same type (obtained by replacing t_i by 0 or 1 for $i \geq 2$) and by another term

$$\sigma\quad(0,t_2,\ldots,t_{p+1}) - \sigma(0,t_2,\ldots,t_{p+1}) - \sigma(\alpha,t_2,\ldots,t_{p+1})$$
$$-\sigma(\alpha+\beta,t_2,\ldots,t_{p+1} + \sigma(\alpha,t_2,\ldots,t_{p+1}) + \sigma(\alpha+\beta,t_2,\ldots,t_{p+1}).$$

This last term is clearly zero   ∎

With the previous definitions the following proposition is trivial.

Proposition 10. *The degenerate p-chains are equivalent to zero for the equivalence relation for p-chains.*   ∎

We denote by $S'_p(M)$ the quotient $\Gamma_p(M)/D_p(M)$ and by $S'^p(M)$ its dual. We can define the integration morphism

$$\int : \Omega^p(M) \to S'^p(M).$$

We can also define the differentiation morphism. In fact we define $\Psi : M^{(D^p)} \times D^p \to R$ by setting $\Psi(\sigma,\underline{e}) = \phi(\sigma^{\underline{e}})$ ($\phi \in S'^p(M)$).

Proposition 11. *There exists a unique p-form $D(\phi$ such that*

$$e_1\cdots e_p(D(\phi)(\sigma) = \phi(\sigma^e).$$

It suffices to show that the morphism Ψ satisfies the conditions of proposition 2 of 4.2. Looking at the identities

$$\sigma^{\alpha}\ i\ {}^{\underline{e}}(t_1,\ldots,t_p) = \sigma(t_1\cdot e_1,\ldots,\alpha\cdot t_i\cdot e_i,\ldots,t_p\cdot e_p),$$
$$(\alpha\ {}_{\cdot i}\ \sigma)^{\underline{e}}(t_1,\ldots,t_p) = \sigma(t_1\cdot e_1,\ldots,\alpha\cdot t_i\cdot e_i,\ldots,t_p\cdot e_p),$$

we have that $\sigma^{\alpha} i^{\underline{e}} = (\alpha \underset{i}{\cdot} \sigma)^{\underline{e}}$. It follows that the multilinearity of Ψ in the variable σ is equivalent to the multilinearity of Ψ in the variable $\underline{e}$. But this last assertion is clear because, if one of the e_i is zero, $\sigma^{\underline{e}}$ is flat and therefore $\Psi(\sigma, \underline{e})$ is multilinear in $\underline{e}$.

The fact that σ is alternating comes from the antisymmetry condition in the definition of degenerate p-chains. ∎

It is easy now to see that D and $\int$ are inverse to each other. The proof is the same as the proof of Theorem 8 with two small exceptions. We need an analogue of Lemmas 6 and 7 for the equivalence relation defined here . For Lemma 6, this is trivial because in $S'^p(M)$ we have finite additivity. For Lemma 7, we have only to remark that the morphism

$$\Psi(d_1, e_1) : (d_2, \ldots, d_P) \to e_1 e_2 \cdots e_p \phi(\sigma(\alpha_1 + d_1 t_1, \ldots, \alpha_p + d_p t_p))$$

vanishes when d_1 or e_1 equals zero : indeed, when $e_1 = 0$ this is obvious and when $d_1 = 0$, the p-cube becomes flat. By the general axiom (KW), $\Psi(d_1, e_1)$ depends only on the product $d_1 \cdot e_1$, and this gives the proof of Lemma 7. Therefore we have the following:

Theorem 12. *D and $\int$ induce inverse isomorphisms between (Ω^p, d) and (S'^p, δ).* ∎

4.6   Commented bibliography

1) The theory of differential forms is mainly developed in [38] by Kock, Reyes and B.Veit(1980). This results from Kock's work [29](1979).

   These questions are also developed in Chapter 4 of Belair's thesis [2](1981) (cf [3]) as likewise in the book of Kock [31](1981).

2) The theory of differential forms leads naturally to de Rahm cohomology, which is analyzed fully and compared to other cohomology theories by Moerdijk and Reyes, for instance in their [50](1983) and [51](1983).

3) Kock has also studied differential forms with value in a Lie group and their link with the more usual differential forms with value

in the Lie algebra of the Lie group ; this work can be found in his [33](1982).

4) a good synthesis of basic facts on differential forms is in the Ph.D. thesis of C. Minguez, prepared under the supervision of Reyes, [47](1985). See also Minguez's [48] and [49](1988).

5) The synthetic version of De Rham's theorem given here comes directly from Felix and Lavendhomme [18](1990).

Chapter 5

Connections

5.1 Connection, covariant derivative and spray

5.1.1 Introduction and definition

An important problem in differential geometry is how one can compare tangent vectors at one point to those at another, at least in the case of nearby points. We ask if, at least during an infinitesimal period of time, it is possible to "transport" a tangent vector in the direction of another tangent vector.

A. Kock and G. Reyes [36] thus describe a connection on an object M as being a datum allowing us to associate to an ordered pair of tangent vectors at a point m a microsquare at m ; in other words to an

141

infinitesimal configuration ofshape :

Figure 3

an infinitesimal configuration of shape :

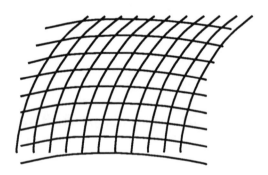

Figure 4

If M is microlinear, the first configuration is an element of $M^{D(2)}$ whereas a microsquare is an element of $M^{D \times D}$. Thus a connection on M is essentially a section of the mapping

$$M^i : M^{D \times D} \to M^{D(2)}$$

on which we impose homogeneity (and thus linearity) conditions ; (here i stands for the canonical injection from $D(2)$ to $D \times D$). We give then the following definition.

Definition 1. *Let M be a microlinear object, $M^D \underset{M}{\times} M^D$ ($\simeq M^{D(2)}$) the set of the ordered pairs (t_1, t_2) of tangent vectors at the same point. A connection on M is a mapping*

$$\nabla : M^D \underset{M}{\times} M^D \to M^{D \times D}$$

such that

$$\nabla(t_1, t_2)(d_1, 0) = t_1(d_1)$$
$$\nabla(t_1, t_2)(0, d_2) = t_2(d_2)$$

$$\nabla(\alpha \cdot t_1, t_2)(d_1, d_2) = \nabla(t_1, t_2)(\alpha \cdot d_1, d_2)$$
$$\nabla(t_1, \alpha \cdot t_2)(d_1, d_2) = \nabla(t_1, t_2)(d_1, \alpha \cdot d_2)$$

for every $(t_1, t_2) \in M^D \underset{M}{\times} M^D$, every d_1, d_2 of D, and every α of R.

Instead of connections we shall also speak of *pointwise connections*, as opposed to the concept of global connection introduced later.

Let us denote by $K : M^{D \times D} \to M^D \times_M M^D$ the composite of M^i with the isomorphism from $M^{D(2)}$ to $M^D \underset{M}{\times} M^D$. The first condition of the preceding definition expresses that ∇ is a section of K. Let us recall that $M^{i_2} : M^{D \times D} \to M^D$ and $M^{i_1} : M^{D \times D} \to M^D$ (where i_1 and i_2 are the injections of D as first or second axis of $D \times D$) are vector fibre bundles : cf. 3.4, Proposition 1. Operations in a fibre of M^{i_2} are denoted by $\underset{1}{+}$ and $\underset{1}{\cdot}$; operations in a fibre of M^{i_1} are denoted by $\underset{2}{+}$ and $\underset{2}{\cdot}$. Thus the two last conditions can be written as follows

$$\nabla(\alpha \cdot t_1, t_2) = \alpha \underset{1}{\cdot} \nabla(t_1, t_2)$$
$$\nabla(t_1, \alpha \cdot t_2) = \alpha \underset{2}{\cdot} \nabla(t_1, t_2)$$

By virtue of (1.2, Proposition 10) we thus have the corresponding linearity conditions.

Proposition 1. *A connection on M is a mapping*

$$\nabla : M^D \underset{M}{\times} M^D \to M^{D \times D}$$

which is a section of K and which is, in two ways, a morphism of vector fibre bundles :

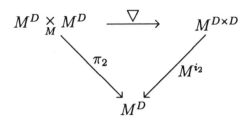

and

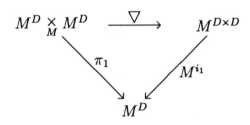

■

Let M be a microlinear object and ∇ a pointwise connection in the above-mentioned sense. We can make ∇ operate globally on vector fields. In the same way as we denote by $\mathcal{X}(M)$ the set of vector fields on M, we shall denote by $\mathcal{X}^{(2)}(M)$ the set of distributions of dimension 2 on M (i.e. the set of microsquare fields, cf. 3.4). To the (pointwise) connection ∇ we associate the mapping, also denoted by ∇ :

$$\nabla : \mathcal{X}(M) \times \mathcal{X}(M) \to \mathcal{X}^{(2)}(M)$$

defined by

$$\nabla(X,Y)_m(d_1, d_2) = \nabla(X_m, Y_m)(d_1, d_2).$$

Let us denote by

$$K : \mathcal{X}^{(2)}(M) \to \mathcal{X}(M) \times \mathcal{X}(M)$$

the mapping obtained by restricting the microsquare fields to their two axes : if $\mathcal{D}$ is a distribution of dimension 2, the two components of $K(\mathcal{D})$ are given by :

$$\begin{aligned} K_1(\mathcal{D})_m(d) &= \mathcal{D}_m(d, 0) \\ K_2(\mathcal{D})_m(d) &= \mathcal{D}_m(0, d). \end{aligned}$$

By construction, ∇ is a section of K satisfying the two homogeneity properties and thus a linearity property. This suggests the following generalization of the connection notion.

Definition 2. *Let M be a microlinear object. A global connection on M is a mapping :*

$$\nabla : \mathcal{X}(M) \times \mathcal{X}(M) \to \mathcal{X}^{(2)}(M)$$

such that

(1) ∇ is a section of the function

$$K : \mathcal{X}^{(2)}(M) \to \mathcal{X}(M) \times \mathcal{X}(M)$$

which is defined by restriction to the axes ;

(2) We have the homogeneity properties

$$\nabla(f \cdot X, Y) = f \underset{1}{\cdot} \nabla(X, Y)$$
$$\nabla(X, f \cdot Y) = f \underset{2}{\cdot} \nabla(X, Y)$$

for every X and Y of $\mathcal{X}(M)$ and every f of R^M.

If one prefers more explicit formulas the conditions on ∇ can be written

(a) $\nabla(X, Y)_m(d_1, 0) = X_m(d_1)$

(b) $\nabla(X, Y)_m(0, d_2) = Y_m(d_2)$

(c) $\nabla(f \cdot X, Y)_m(d_1, d_2) = \nabla(X, Y)_m(f(m) \cdot d_1, d_2)$

(d) $\nabla(X, f \cdot Y)_m(d_1, d_2) = \nabla(X, Y)_m(d_1, f(m) \cdot d_2)$.

The remarks that precede the definition immediately give :

Proposition 2. *A pointwise connection on M induces a global connection.* ∎

Note that the conditions (2) (made more explicit by (c) and (d)) are linearity conditions for structures of R^M-modules which imply the corresponding conditions for R-module structures. But we don't have the converse. Let us consider, for instance, for X and Y in $\mathcal{X}(M)$, the field of microsquares $Y \star X$ given by

$$(Y \star X)_m(d_1, d_2) = Y(X(m, d_1), d_2).$$

We have :

$$
\begin{array}{lll}
(a) & (Y \star X)_m(d_1, 0) &= Y(X(m, d_1), 0) = X_m(d_1) \,; \\
(b) & (Y \star X)_m(0, d_2) &= Y(X(m, 0), d_2) = Y_m(d_2) \,; \\
(c) & (Y \star fX)_m(d_1, d_2) &= Y((fX)(m, d_1), d_2) \\
& &= Y(X(m, f(m) \cdot d_1), d_2) \\
& &= (Y \star X)_m(f(m) \cdot d_1, d_2) \,;
\end{array}
$$

but in general we do not have the identity (d), although we have for $\alpha \in R$:

$$
\begin{aligned}
(\alpha Y \star X)_m(d_1, d_2) &= (\alpha Y)(X(m, d_1), d_2) \\
&= Y(X(m, d_1), \alpha \cdot d_2) \\
&= (Y \star X)_m(d_1, \alpha \cdot d_2).
\end{aligned}
$$

We shall subsequently need the formula giving $fY \star X$:

Proposition 3. *We have, for every $f \in R^M$,*

$$
fY \star X = f \underset{2}{\cdot} (Y \star X) \dot{+} X(f) \cdot Y.
$$

Proof: We first have

$$
\begin{aligned}
(fY \star X)_m(d_1, d_2) &= (fY)(X(m, d_1), d_2) \\
&= (f(X(m, d_1)) \cdot Y_{X(m, d_1)})(d_2) \\
&= Y(X(m, d_1), f(X(m, d_1)) \cdot d_2).
\end{aligned}
$$

This gives by definition of $X(f)$ (cfr. 3.3.1) :

$$
Y(X(m, d_1), f(m) \cdot d_2 + d_1 \cdot d_2 \cdot X(f)(m)).
$$

Now, let us recall (cf. 3.4, second lemma) that the diagram

$$
\begin{array}{ccc}
1 & \longrightarrow & D \\
\downarrow & & \downarrow{\scriptstyle \epsilon} \\
D \times D & \xrightarrow{\ \varphi\ } & (D \times D) \vee D
\end{array}
$$

(where $\epsilon(d) = (0,0,d)$ and $\varphi(d_1, d_2) = (d_1, d_2, 0)$) is a quasi-colimit, which is what allows us to define the sum $\dot{+}$ of a microsquare and a vector (3.4, Proposition 4).

For each m in M we define

$$\mu_m(d_1, d_2, e) = Y(X(m, d_1), f(m)d_2 + e \cdot X(f)(m)).$$

On one hand we have

$$
\begin{aligned}
\mu_m(0, 0, e) &= Y(m, e.X(f)(m)) \\
&= (X(f).Y)_m(e) \\
\mu_m(d_1, d_2, 0) &= Y(X(m, d_1), f(m)d_2) \\
&= (f \underset{2}{\cdot} (Y \star X))_m(d_1, d_2)
\end{aligned}
$$

and on the other hand

$$\mu_m(d_1, d_2, d_1 \cdot d_2) = (fY \star X)_m(d_1, d_2).$$

Thus we have the announced formula. ∎

5.1.2 Covariant derivative

If a connection allows to compare vectors at points near each other, it is natural to ask whether it would not allow us to differentiate one vector field with respect to another.

In order to study this, let us begin by defining a notion of derivative.

Definition 3. *Let M be a microlinear object. A covariant derivation is a mapping*

$$\nabla : \mathcal{X}(M) \times \mathcal{X}(M) \to \mathcal{X}(M) : (X, Y) \rightsquigarrow \nabla_X Y$$

satisfying the identities :

1) $\nabla_{f \cdot X} Y = f \cdot \nabla_X Y$

2) $\nabla_X(f \cdot Y) = f \cdot \nabla_X Y + X(f) \cdot Y$

for every f of R^M.

The second identity does indeed suggest an algebraic law of differentiation.

Notice that we have not explicitly imposed R-bilinearity because it follows from the two axioms given. Actually, the first one immediatly gives $\nabla_{\alpha \cdot X} Y = \alpha \cdot \nabla_X Y$ for every α of R, but the second one gives also $\nabla_X (\alpha \cdot Y) = \alpha \cdot \nabla_X Y$ because $X(\alpha)$ is null. This bihomogeneity furnishes as usual bilinearity (1.2, Prop.10). Thus we have in particular,

$$\nabla_{X_1 + X_2}(Y) = \nabla_{X_1}(Y) + \nabla_{X_2}(Y)$$

and

$$\nabla_X (Y_1 + Y_2) = \nabla_X Y_1 + \nabla_X Y_2.$$

If $\nabla : \mathcal{X}(M) \times \mathcal{X}(M) \to \mathcal{X}^{(2)}(M)$ is a global connection, the covariant derivative that we are going to associate to it measures, for every X and Y of $\mathcal{X}(M)$, the difference that exists between $\nabla(X, Y)$ and the microsquare field given directly by $Y \star X$. More precisely $\nabla(X, Y)$ and $Y \star X$ coincide on the axes and thus we can consider the strong difference (cfr. 3.4)

$$Y \star X \mathbin{\dot{-}} \nabla(X, Y).$$

Proposition 4. *If ∇ is a global connection, we define a covariant derivation by putting*

$$\nabla_X Y = Y \star X \mathbin{\dot{-}} \nabla(X, Y).$$

We have so a bijection from the set of global connections to the set of covariant derivations.

Proof: Let ∇ be a global connection. Let us recall

$$
\begin{aligned}
Y \star f \cdot X &= f \mathbin{\underset{i}{\cdot}} (Y \star X) \\
\nabla(f \cdot X, Y) &= f \mathbin{\underset{i}{\cdot}} \nabla(X, Y).
\end{aligned}
$$

Thus we have

$$
\begin{aligned}
\nabla_{f \cdot X} Y &= Y \star f X \mathbin{\dot{-}} \nabla(f X, Y) \\
&= f \mathbin{\underset{i}{\cdot}} (Y \star X) \mathbin{\dot{-}} f \mathbin{\underset{i}{\cdot}} \nabla(X, Y) \\
&= f \cdot (Y \star X \mathbin{\dot{-}} \nabla(X, Y)) \\
&= f \cdot \nabla_X Y.
\end{aligned}
$$

(by virtue of 3.4, Proposition 5).

We have the Koszul law since

$$\nabla(X, f \cdot Y) = f \underset{2}{.} \nabla(X, Y)$$

by the very definition of global connections, and

$$(fY) \star X = f \underset{2}{.} (Y \star X) \dotplus X(f).Y$$

by Proposition 3 of the previous section.

$$\begin{aligned}
\nabla_X(f \cdot Y) &= (fY) \star X \dot{-} \nabla(X, fY) \\
&= (f \underset{2}{.} (Y \star X) \dotplus X(f) \cdot Y) \dot{-} f \underset{2}{.} \nabla(X, Y)
\end{aligned}$$

which gives, by virtue of the structure of affine spaces (3.4, Proposition 4),

$$(f \underset{2}{.} (Y \star X) \dot{-} f \underset{2}{.} \nabla(X, Y)) + X(f) \cdot Y$$

or, by (3.4, Proposition 5) :

$$f \cdot \nabla_X + X(f) \cdot Y.$$

Conversely, if $\nabla_X Y$ is a covariant derivation, we put

$$\nabla(X, Y) = Y \star X \dotplus (-\nabla_X Y).$$

Let us verify that we do have a global connection. Since we have

$$\nabla(X, Y)_m(d_1, d_2) = u_m(d_1, d_2, d_1 \cdot d_2)$$

where $u_M : (D \times D) \vee D$ is characterized by

$$\begin{aligned}
u_m(d_1, d_2, 0) &= (Y \star X)_m(d_1, d_2) \\
&= Y(X(m, d_1), d_2)
\end{aligned}$$

and

$$u_m(0, 0, d) = (\nabla_X Y)_m(-d),$$

we have

$$
\begin{aligned}
\nabla(X,Y)_m(d_1,0) &= u_m(d_1,0,0) \\
&= Y(X(m,d_1),0) \\
&= X_m(d_1). \\
\nabla(X,Y)_m(0,d_2) &= u_m(0,d_2,0) \\
&= Y(X(m,0),d_2) \\
&= Y_m(d_2).
\end{aligned}
$$

On the other hand, by virtue of (3.4, Prop. 5),

$$
\begin{aligned}
\nabla(fX,Y) &= (Y \star fX)\dotplus(-\nabla_{fX}Y) \\
&= f \underset{1}{\,}(Y \star X)\dotplus f.(-\nabla_X Y) \\
&= f \underset{1}{\,} \nabla(X,Y)
\end{aligned}
$$

and

$$
\nabla(X,fY) = (fY \star X)\dotplus(-\nabla_X fY)
$$

hence, by the Koszul law and Proposition 3,

$$
\begin{aligned}
\nabla(X,fY) &= (f \underset{2}{\,}(Y \star X)\dotplus X(f)\cdot Y)\dotplus - (f\cdot(\nabla_X Y) + X(f)\cdot Y) \\
&= f \underset{2}{\,}(Y \star X)\dotplus f\cdot(-\nabla_X Y) \\
&= f \underset{2}{\,} \nabla(X,Y).
\end{aligned}
$$

Elementary computations involving properties of affine spaces then show that the two constructions are inverse to each other.   ∎

If the connection ∇ is not only global, but pointwise, we can improve a little bit the previous construction by defining a notion of the covariant derivative of a field Y with respect to a tangent vector. Let $t \in T_m M$ be a tangent vector and Y a vector field on M. We define the tangent vector

$$
\nabla_t Y \in T_m M
$$

by

$$
\nabla_t Y = Y \star t \dotminus \nabla(t,Y_m)
$$

where $Y \star t$ is a microsquare

$$
(Y \star t)(d_1,d_2) = Y(t(d_1),d_2).
$$

Thus we have a mapping from $T_m M \times \mathcal{X}(M)$ into $T_m M$ having the properties

1) $\nabla_{\alpha \cdot t} Y = \alpha \cdot \nabla_T Y$ for every $\alpha \in R$

2) $\nabla_t (f \cdot Y) = f(m) \cdot \nabla_t Y + t(f) \cdot Y_m$

for every f of R^M and where $t(f)$ is characterized by

$$f(t(d)) = f(m) + d \cdot t(f).$$

5.1.3 Sprays

If a pointwise connection allows to complete the infinitesimal configuration formed by two tangent vector to a microsquare, it allows it *a fortiori* if the vectors coincide. If the vectors differ slightly, the transformation is intuitively suggested by the following figure :

Figure 5

but if the two vectors coincide, we get this figure :

Figure 6

Thus we can expect a pointwise connection to determine a mapping

$$\sigma : M^D \to M^{D_2}$$

(where D_2 is the small object $\{\delta \in R \mid \delta^3 = 0\}$).

     This mapping σ extends a tangent vector, or "1-arc", into a "2-arc". Hence the name of spray given to this extension.

Definition 4. *A spray on M is a mapping*

$$\sigma : M^D \to M^{D_2}$$

such that :

(1) $\sigma(t)(d) = t(d)$ *for every t of M^D and every d of D;*

(2) $\sigma(\alpha \cdot t)(\delta) = \sigma(t)(\alpha \cdot \delta)$ *for every t of M^D, every α of R and every δ of D_2.*

Every pointwise connection determines a spray :

Proposition 5. *Let M be a microlinear object and let ∇ be a pointwise connection on M. The formula*

$$\sigma(t)(d_1 + d_2) = \nabla(t, t)(d_1, d_2)$$

defines a spray on M.

Proof: Let ∇ be a connection. As the diagram

$$D \underset{i_2}{\overset{i_1}{\rightrightarrows}} D \times D \xrightarrow{\ +\ } D_2$$

is a quasi-colimit (cfr. 2.2) and as

$$\nabla(t, t)(d, 0) = t(d) = \nabla(t, t)(0, d),$$

the function $\nabla(t, t) : D \times D \to M$ factorizes in a unique way by D_2 (because M is microlinear). Thus we do have a unique mapping $\sigma(t) : D_2 \to M$ such that

$$\sigma(t)(d_1 + d_2) = \nabla(t, t)(d_1, d_2).$$

It remains to observe that it really is a spray.

(a) $\sigma(t)(d) = \nabla(t, t)(d, 0) = t(d)$ for every t of d and

(b)

$$\begin{aligned}
\sigma(\alpha \cdot t)(d_1 + d_2) &= \nabla(\alpha \cdot t, \alpha \cdot t)(d_1, d_2) \\
&= \nabla(t, t)(\alpha \cdot d_1, \alpha \cdot d_2) \\
&= \sigma(t)(\alpha \cdot (d_1 + d_2))
\end{aligned}$$

and thus, as R thinks that addition is surjective (2.2, Proposition 2), for every δ of D_2, $\sigma(\alpha \cdot t)(\delta) = \sigma(t)(\alpha \cdot \delta)$. ∎

We wish to show now that, conversely, a spray determines a connection.

We shall use the following easy lemma.

Lemma 6. *The diagram*

$$D_2 \times D_2 \times D \times D \underset{v}{\overset{u}{\rightrightarrows}} D_2 \times D \times D \xrightarrow{\quad w \quad} D \times D$$

where

$$
\begin{aligned}
u(\delta_1, \delta_2, d_1, d_2) &= (\delta_1 \cdot \delta_2, d_1, d_2), \\
v(\delta_1, \delta_2, d_1, d_2) &= (\delta_2, \delta_1 \cdot d_1, \delta_1 \cdot d_2), \\
w(\delta, d_1, d_2) &= (\delta \cdot d_1, \delta \cdot d_2),
\end{aligned}
$$

is a quasi-colimit.

Proof: Being a matter of small objects, it suffices to verify that R perceives the diagram as a colimit (2.2, Prop. 9). Let $f : D_2 \times D \times D \to R$ equalize u and v. We have

$$
\begin{aligned}
f(\delta, d_1, d_2) = \ & a_0 + a_1 d_1 + a_2 d_2 + a_3 d_1 d_2 \\
+ \ & b_1 \delta + b_2 d_1 \delta + b_3 d_2 \delta + b_4 d_1 d_2 \delta \\
+ \ & c_1 \delta^2 + c_2 d_1 \delta^2 + c_3 d_2 \delta^2 + c_4 d_1 d_2 \delta^2,
\end{aligned}
$$

with $f(\alpha \cdot \delta, d_1, d_2) = f(\delta, \alpha \cdot d_1, \alpha \cdot d_2)$ for every α of D_2.

(a) Put $\alpha = 0$. We get for every d_1, d_2 and δ :

$$a_0 + a_1 d_1 + a_2 d_2 + a_3 d_1 d_2 = a_0 + b_1 \delta + c_1 \delta^2$$

and thus $a_1 = a_2 = a_3 = b_1 = c_1 = 0$.

(b) Let us then put $\alpha = d_1$; we get

$$a_0 + b_3 d_2 d_1 \delta = a_0 + b_3 d_1 d_2 \delta + c_3 d_1 d_2 \delta^2$$

and thus $c_3 = 0$. Putting similarly $\alpha = d_2$, we find $c_2 = 0$.

(c) Finally with $\alpha = \delta$, we get

$$a_0 + b_2 d_1 \delta^2 + b_3 d_2 \delta^2 + b_4 d_1 d_2 \delta^2 = a_0 + b_2 d_1 \delta^2 + b_3 d_2 \delta^2$$

and thus $b_4 = 0$. So we have

$$f(\delta, d_1, d_2) = a_0 + b_2 d_1 \delta + b_3 d_2 \delta + c_4 d_1 d_2 \delta^2.$$

Putting $g : D \times D \to R : (d_1, d_2) \rightsquigarrow a_0 + b_2 d_1 + b_3 d_2 + c_4 d_1 d_2$ we see that g is the only function such that

$$f(\delta, d_1, d_2) = g(\delta \cdot d_1, \delta \cdot d_2).$$

■

Let now σ be a spray on M. Let t_1 and t_2 be two tangent vectors at the same point. Consider the function f from $D_2 \times D \times D$ to M given by

$$f(\delta, d_1, d_1) = \sigma(d_1 t_1 + d_2 t_2)(\delta).$$

For every α in D_2 we have

$$\begin{aligned}
f(\alpha \cdot \delta, d_1, d_2) &= \sigma(d_1 t_1 + d_2 t_2)(\alpha.\delta) \\
&= \sigma(\alpha d_1 t_1 + \alpha d_2 t_2)(\delta) \\
&= f(\delta, \alpha \cdot d_1, \alpha \cdot d_2)
\end{aligned}$$

and thus, taking into account the previous lemma and the microlinearity of M, there exists one and only one function g from $D \times D$ to M such that

$$f(\delta, d_1, d_2) = g(\delta d_1, \delta d_2).$$

We denote this function g by $\nabla_\sigma(t_1, t_2)$.

Proposition 7. *Let M be microlinear and let σ be a spray on M. The function ∇_σ described above and characterized by*

$$\nabla_\sigma(t_1, t_2)(\delta \cdot d_1, \delta \cdot d_2) = \sigma(d_1 t_1 + d_2 t_2)(\delta)$$

($\delta \in D_2, d_1, d_2 \in D$), is a connection on M.

Proof: We have first

$$
\begin{aligned}
\nabla_\sigma(t_1, t_2)(\delta \cdot d_1, 0) &= \sigma(d_1 t_1)(\delta) \\
&= \sigma(t_1)(d_1 \delta) \\
&= t_1(d_1 \delta)
\end{aligned}
$$

and in the same way $\nabla_\sigma(t_1, t_2)(0, \delta \cdot d_2) = t_2(d_2 \delta)$. Thus as R (and so M) believes that the multiplication $D_2 \times D \xrightarrow{m} D$ is surjective, we do have

$$
\begin{aligned}
\nabla_\sigma(t_1, t_2)(d, 0) &= t_1(d) \\
\nabla_\sigma(t_1, t_2)(0, d) &= t_2(d).
\end{aligned}
$$

On the other hand we have

$$
\begin{aligned}
\nabla_\sigma(\alpha t_1, \beta t_2)(\delta d_1, \delta d_2) &= \sigma(\alpha d_1 t_1 + \beta d_2 t_2)(\delta) \\
&= \nabla_\sigma(t_1, t_2)(\delta \alpha d_1, \delta \beta d_2)
\end{aligned}
$$

and thus the desired homogeneity. ∎

The connection ∇_σ possesses a further property of symmetry.

Definition 5. *We say that a connection ∇ is a symmetric connection if*

$$
\nabla(t_1, t_2)(d_1, d_2) = \nabla(t_2, t_1)(d_2, d_1).
$$

Proposition 8. *For every spray σ, the connection ∇_σ is symmetric.*

We now have the following result, sometimes called the Ambrose-Palais-Singer theorem.

Proposition 9. *The indicated constructions in Propositions 5 and 7 establish a bijection between the set of sprays and the set of symmetrical pointwise connections.*

Proof:
(a) Let σ be a spray, ∇ the corresponding symmetric connection described in Proposition 7 and $\overline{\sigma}$ the spray corresponding to ∇ described in Proposition 5. We have for every δ of D_2 and for every d_1, d_2 of D

$$
\begin{aligned}
\overline{\sigma}(t)(\delta(d_1 + d_2)) &= \nabla(t, t)(\delta d_1, \delta d_2) \\
&= \sigma(d_1 t + d_2 t)(\delta) \\
&= \sigma(t)(\delta(d_1 + d_2)).
\end{aligned}
$$

As R, and thus M, perceives the addition

$$D \times D \xrightarrow{+} D_2$$

and the multiplication

$$D_2 \times D_2 \xrightarrow{m} D_2$$

as surjective, we have $\overline{\sigma}(t) = \sigma(t)$.

(b) Conversely, if ∇ is a symmetric connection, σ the corresponding spray and $\overline{\nabla}$ the corresponding connection to σ, we must show that $\overline{\nabla} = \nabla$. By virtue of Lemma 6, it suffices to verify that for every δ in D_2 and every $d_1, d_2 \in D$,

$$\overline{\nabla}(t_1, t_2)(\delta \cdot d_1, \delta d_2) = \nabla(t_1, t_2)(\delta \cdot d_1, \delta \cdot d_2).$$

As R, and thus M, believes that the addition

$$D \times D \xrightarrow{+} D_2$$

is surjective, it suffices to verify that for every d_3 and d_4 in D :

$$\overline{\nabla}(t_1, t_2)((d_3 + d_4)d_1, (d_3 + d_4)d_2)$$

$$= \nabla(t_1, t_2)((d_3 + d_4) \cdot d_1, (d_3 + d_4) \cdot d_2). \tag{1}$$

By definition of $\overline{\nabla}$, the lefthand side is equivalent to

$$\sigma(d_1 t_1 + d_2 t_2)(d_3 + d_4) = \nabla(d_1 t_1 + d_2 t_2, d_1 t_1 + d_2 t_2)(d_3, d_4).$$

In order to prove (1), we are actually going to prove that :

$$\nabla(d_1 t_1 + d_2 t_2, d_1' t_1 + d_2' t_2)(d_3, d_4)$$

$$= \nabla(t_1, t_2)(d_3 d_1 + d_4 d_1', d_3 d_2 + d_4 d_2') \tag{2}$$

for every (d_1, d_1') in $D(2)$ and (d_2, d_2') in $D(2)$. By putting $d_1' = d_1$ and $d_2' = d_2$ we do find (1). To prove (2), it suffices (cf. 2.3, Proposition 2) to verify for $d_1 = 0$ and for $d_1' = 0$; and for each case, to verify it for $d_2 = 0$ and for $d_2' = 0$. Hence the following four small verifications :

(i) Put $d_1' = d_2' = 0$; the equality to be proved becomes

$$\nabla(d_1 t_1 + d_2 t_2, 0)(d_3, d_4) = \nabla(t_1, t_2)(d_3 d_1, d_3 d_2). \qquad (2.i)$$

The lefthand side of (2.i) is

$$(d_1 t_1 + d_2 t_2)(d_3).$$

To compute the righthand side of (2.i), we observe that

$$\nabla(t_1, t_2)(d_3 d_1, 0) = t_1(d_3 d_1) = (d_1 t_1)(d_3)$$

$$\nabla(t_1, t_2)(0, d_3 d_2) = t_2(d_3 d_2) = (d_2 t_2)(d_3)$$

and thus that

$$\nabla(t_1, t_2)(d_3 d_1, d_3 d_2) = (d_1 t_1 + d_2 t_2)(d_3).$$

(ii) Put $d_1' = d_2 = 0$; the equality to be proved becomes

$$\nabla(d_1 t_1, d_2' t_2)(d_3, d_4) = \nabla(t_1, t_2)(d_3 d_1, d_4 d_2'). \qquad (2.ii)$$

which follows from the definition of the connection.

(iii) Put $d_1 = d_2' = 0$; the equality to be proved becomes

$$\nabla(d_2 t_2, d_1' t_1)(d_3, d_4) = \nabla(t_1, t_2)(d_4 d_1', d_3 d_2). \qquad (2.iii)$$

This follows from the linearity properties and the symmetry of ∇.

(iv) Finally, put $d_1 = d_2 = 0$; the equality to be proved this time is :

$$\nabla(0, d_1' t_1 + d_2' t_2)(d_3, d_4) = \nabla(t_1, t_2)(d_4 d_1', d_4 d_2') \qquad (2.iv)$$

which, given the symmetry of ∇, reduces to (2.i).

Thus the proposition is proved.

∎

5.2 Vertical and horizontal microsquares

5.2.1 Vertical-horizontal decomposition

A microsquare $\gamma : D \times D \to M$ may of course be considered as a tangent vector to M^D :

$$\tau : D \to M^D.$$

The correspondence is described by

$$\tau(d_1)(d_2) = \gamma(d_1, d_2);$$

and follows from the obvious bijection

$$M^{D \times D} \simeq (M^D)^D.$$

We say that the microsquare γ is *vertical* if the vector τ, tangent to M^D and corresponding to γ, is in fact tangent to the fibre $T_m(M)$ (where $m = \gamma(0,0)$ is the center of γ). Let us denote, as we did before, by $K : M^{D \times D} \to M^D \times M^D : \gamma \rightsquigarrow K(\gamma) = (t_1, t_2)$, the restriction function at the axes (i.e. $t_1(d) = \gamma(d, 0)$ and $t_2(d) = \gamma(0, d)$). To say that γ is vertical means that $\tau(d)(0) = m$ for every d, i.e. that $\gamma(d, 0) = m$, in other words that t_1 is the null vector. The contact point of τ i.e. $\tau(0)$, is then the element t_2 in $T_m(M)$.

Note also that, as $T_m(M)$ is a Euclidean R-module (3.1, Prop. 2), the function $\tau : D \to T_m(M)$ corresponding to the vertical microsquare γ, is written in a unique way as a first degree function : there exists a unique vector $s \in T_m(M)$ such that :

$$\tau(d) = t_2 + d \cdot s$$

for every d of D. We say that s is the *deviation* of the vertical microsquare γ. We say also that t_2 is the *axis* of γ.

The set of vertical microsquares centered at m will be denoted by V_m :

$$V_m \simeq (T_m M)^D.$$

For a given t_2 in $T_m(M)$, we denote by $V_m(t_2)$ the set of vertical microsquares the axis of which is t_2 :

$$V_m(t_2) \simeq T_{t_2}(T_m M).$$

Note that $V_m(t_2)$ is a sub-R-module of the fibre above t_2 of the vector bundle

$$M^{i_2} : M^{D \times D} \to M^D.$$

The function that associates to $\tau \in V_m(t_2)$ its deviation s furnishes an R-module isomorphism from $V_m(t_2)$ to $T_m(M)$.

Let us suppose now that we have a connection ∇ on M. We say that a microsquare γ is horizontal if it is in the image of ∇. The set of horizontal microsquares in M will be denoted by H_m; for $t_2 \in T_m(M)$, the fibre of H_m over t_2 is denoted by $H_m(t_2)$. It is clearly an R-module.

We described in this way two vector sub-bundles of

$$M^{i_2} : M^{D \times D} \to M^D$$

that we shall write

$$V_M : V(M) \to M^D$$
$$H_M : H(M) \to M^D;$$

they are fibres of the form $V_m(t_2)$ and $H_m(t_2)$ respectively.

If one has two vector sub-bundles A and B of a vector bundle $E \to X$, the *Whitney sum* of A and B is obtained by summing, for every x of X, A_x and B_x (as sub-R modules of E_x). We denote this Whitney sum by $A \oplus B$.

Proposition 1. *The vector bundle*

$$M^{i_2} : M^{D \times D} \to M^D$$

is the Whitney sum of the bundles V_M and H_M.

Proof: Let $t_2 \in T_m(M) \subseteq M^D$ and let γ be in the fibre bundle of M^{i_2} over t_2. The horizontal and vertical components of γ are defined by

$$
\begin{aligned}
H(\gamma) &= \nabla K(\gamma) \in H_m(t_2) \\
V(\gamma) &= \gamma \underset{1}{\cdot} \nabla K(\gamma) \in V_m(t_2).
\end{aligned}
$$

In this way we have the required decomposition

$$M_{1,t_2} = M^{i_2}(t_2) \underset{<V,H>}{\overset{\underset{1}{+}}{\rightleftarrows}} V_m(t_2) \oplus H_m(t_2).$$

i) We immediately have for every $\gamma \in M^{i_2}(t_2)$

$$V(\gamma) \underset{1}{+} H(\gamma) = \gamma$$

ii) Let $A \in V_m(t_2)$ and $B \in H_m(t_2)$. We have

$$H(A \underset{1}{+} B) = \nabla K(A \underset{1}{+} B).$$

But as the restriction of A to the first axis of $D \times D$ is null and as the restrictions of A and B to the second axis of $D \times D$ are t_2, we have :

$$K(A \underset{1}{+} B) = K(B) = (t_1, t_2)$$

and thus

$$\nabla K(A \underset{1}{+} B) = \nabla(t_1, t_2) = B$$

this last equality following from the fact that B is horizontal. We also have

$$\begin{aligned} V(A \underset{1}{+} B) &= (A \underset{1}{+} B) \underset{1}{\dot{-}} \nabla K(A \underset{1}{+} B) \\ &= (A \underset{1}{+} B) \underset{1}{\dot{-}} B = A. \end{aligned}$$

iii) Finally if $A \in V_m(t_2) \cap H_m(t_2)$,

$$A = \nabla K(A) = \nabla(0, t_2)$$

which is the null microsquare of the fibre. Thus we do have

$$M^{i_2} \simeq V_M \oplus H_M.$$

■

We have a partial converse to Proposition 1. Suppose we have a decomposition of the vector bundle

$$M^{i_2} : M^{D \times D} \to M^D$$

as a Whitney sum $V_M \oplus H_M$ of sub-bundles of M^{i_2} where V_M is the vector bundle of vertical microsquares. We denote by V and H the resulting map from $M^{D \times D}$ to V_M and H_M respectively. We say that the decomposition is compatible with the structure $\underset{2}{\cdot}$ if $H(\alpha \underset{2}{\cdot} \gamma) = \alpha \underset{2}{\cdot} H(\gamma)$ (for $\alpha \in R$ and $\gamma \in M^{D \times D}$).

Proposition 2. *Let (M, ∇_1) be a microlinear object equipped with an arbitray connection ∇_1. Suppose the vector bundle*

$$\pi_2 : \mathcal{X}^{(2)}(M) \to \mathcal{X}(M)$$

decomposes itself in a Whitney sum as above, and this decomposition is compatible with the structure $\underset{2}{\cdot}$ Then there exists a connection ∇ such that H_M is the vector bundle of horizontal microsquares for ∇.

Proof: Let $t_1, t_2 \in T_m M$. We put

$$\nabla(t_1, t_2) = H(\nabla_1(t_1, t_2)).$$

We have the compatibilities of H and ∇_1 with the two stuctures $\underset{1}{\cdot}$ and $\underset{2}{\cdot}$ and thus we have the two linearity conditions on ∇. As $\nabla(t_1, t_2) = H(\nabla_1(t_1, t_2))$ we have $K(\nabla(t_1, t_2)) = K(\nabla_1(t_1, t_2)) = (t_1, t_2)$ and the proof is complete. ∎

5.2.2 Connecting mappings or connection forms

Let $\gamma \in M^{D \times D} \simeq (M^D)^D$ be a microsquare. If ∇ is a (pointwise) connection on M, we can consider the gap existing between the microsquare γ and the parallel prolongation $\nabla(t_1, t_2)$ of $(t_1, t_2) = K(\gamma)$. A way of seeing this gap would be to consider the vertical component of γ

$$V(\gamma) = \gamma \underset{1}{\cdot} \nabla K_\gamma.$$

Actually, t_2 being fixed, $V(\gamma)$ is well described by its deviation, which we denote by $C(\gamma)$.

As a matter of fact, this deviation $C(\gamma)$ can also be expressed by the strong difference.

Proposition 3. *For every $\gamma \in M^{D \times D}$, we have*

$$C(\gamma) = \gamma \overset{\cdot}{-} \nabla(t_1, t_2).$$

Proof: By the very definition of the strong difference, we have

$$(\gamma \overset{\cdot}{-} \nabla(t_1, t_2))(e) = \ell(0, 0, e)$$

where $\ell : (D \times D) \vee D \to M$ is characterized by

$$\begin{aligned} \ell(d_1, d_2, 0) &= \nabla(t_1, t_2)(d_1, d_2) \\ \ell(d_1, d_2, d_1 \cdot d_2) &= \gamma(d_1, d_2). \end{aligned}$$

Let us compute $\ell(0, d_2, d_1 \cdot d_2)$. In order to do so, we consider $\ell(0, d_2, d_1 \cdot d_2')$ for $(d_2, d_2') \in D(2)$. We have

$$\begin{aligned} \ell(0, d_2, 0) &= \nabla(t_1, t_2)(0, d_2) = t_2(d_2) \\ \ell(0, 0, d_1 \cdot d_2') &= (\gamma \dot{-} \nabla(t_1, t_2))(d_1 \cdot d_2') \\ &= d_1 \cdot (\gamma \dot{-} \nabla(t_1, t_1))(d_2'), \end{aligned}$$

and so

$$\ell(0, d_2, d_1 \cdot d_2) = (t_2 + d_1(\gamma \dot{-} \nabla(t_1, t_2)))(d_2).$$

Put

$$A(d_1, d_2) = (t_2 + d_1(\gamma \dot{-} \nabla(t_1, t_2)))(d_2).$$

We consider $\ell(d_1, d_2, d_1' \cdot d_2)$ for $(d_1, d_1') \in D(2)$. By setting $d_1' = 0$, we get $\ell(d_1, d_2, 0) = \nabla(t_1, t_2)(d_1, d_2)$, whereas if we write $d_1 = 0$ we get

$$\ell(0, d_2, d_1' \cdot d_2) = A(d_1', d_2).$$

But then, by setting $d_1' = d_1$, we get the $\underset{1}{+}$ sum of these microsquares :

$$\ell(d_1, d_2, d_1 \cdot d_2) = (\nabla(t_1, t_2) \underset{1}{+} A)(d_1, d_2).$$

But the left hand side is $\gamma(d_1, d_2)$, thus

$$\nabla(t_1, t_2) \underset{1}{+} A = \gamma$$

or

$$\begin{aligned} t_2 + d_1(\gamma \dot{-} \nabla(t_1, t_2)) &= (\gamma \dot{-} \nabla(t_1, t_2))(d_1, \cdot) \\ &= t_2 + d_1 c(\gamma), \end{aligned}$$

and hence the claim.   ∎

We make then the following definition :

Definition 12. *Let ∇ be a pointwise connection, we call*

$$C : M^{D \times D} \to M^D$$

defined by $C(\gamma) = \gamma \dot{-} \nabla K \gamma$ a connecting mapping or connection form. In the same way, if ∇ is global, we define

$$C : \mathcal{X}^{(2)}(M) \to \mathcal{X}(M)$$

by $C(\gamma)_m = \gamma_m \dot{-} (\nabla K \gamma)_m$ for every m of M, and we call it the global connecting mapping.

Proposition 4. *The connecting mapping is R-linear for the two structures of vector fibre bundles of $M^{D \times D}$ on M^D. In the same way for global connecting mappings we have R^M-linearity (in both ways).*

Proof: Being a matter of trivial computation, we shall merely verify one of the two R^M-linearity properties. Let $\gamma \in \mathcal{X}^{(2)}(M)$ with $K(\gamma) = (X, Y)$. We have for every f of R^M

$$
\begin{aligned}
C(f \underset{i}{\cdot} \gamma)_m &= (f \underset{i}{\cdot} \gamma)_m \dot{-} (\nabla K(f \underset{i}{\cdot} \gamma))_m \\
&= f(m) \underset{i}{\cdot} \gamma_m \dot{-} \nabla(f \cdot X, Y)_M \\
&= f(m) \underset{i}{\cdot} \gamma_m \dot{-} f(m) \underset{i}{\cdot} \nabla(X, Y)_m \\
&= f(m) \cdot (\gamma_m \dot{-} \nabla(X, Y)_m)
\end{aligned}
$$

(by 3.4, Proposition 5).

$$= (f \cdot C(\gamma))_m.$$

∎

As an application of this definition let us point out the translation of Proposition 4 of 5.1:

Proposition 5. *If ∇ is a global connection, we have*

$$\nabla_X Y = C(Y \star X).$$

∎

5.2.3   Parallel transport

Let (t, e) be a marked microsegment at $m = t(0)$; in other words let $t \in T_m M$ and $e \in D$. We wish to specify here the notion of (infinitesimal) *parallel transport* associated to a connection ∇, suggested in section 5.1.

We consider the following mappings

$$p_{(t,e)} : T_m M \to T_{t(e)} M$$

defined by

$$p_{(t,e)}(t')(d) = \nabla(t, t')(e, d),$$

and

$$q_{(t,e)} : T_{t(e)} M \to T_m M$$

defined by

$$q_{(t,e)}(t_1)(d) = \nabla(\nabla(t, t)(e, \cdot), t_1)(-e, d).$$

Proposition 6. *The mappings $p_{(t,e)}$ and $q_{(t,e)}$ are inverse isomorphisms between $T_m M$ and $T_{t(e)} M$.*

Proof:

(a) Let us first observe that $p_{(t,e)}(t')$ really is in $T_{t(e)} M$ because

$$p_{(t,e)}(t')(0) = \nabla(t, t')(e, 0) = t(e).$$

Next, we observe that $q_{(t,e)}$ is well defined : first $\nabla(t, t)(e, \cdot)$ is in $T_{t(e)} M$ as is t_1, and then

$$\begin{aligned} q_{(t,e)}(t_1)(0) &= \nabla(\nabla(t, t)(e, \cdot), t_1)(-e, 0) \\ &= \nabla(t, t)(e, -e) \end{aligned}$$

and as

$$\nabla(t, t)(e, 0) = t(e) = \nabla(t, t)(0, e),$$

we have $\nabla(t, t)(e, -e) = (t - t)(e) = 0(e) = m$.

(b) We show that $q_{(t,e)}(p_{(t,e)}(t')) = t'$. Let $(e, e') \in D(2)$. Consider

$$L(e, e') = \nabla(\nabla(t, t)(e, \cdot), \nabla(t, t')(e, \cdot))(e', \cdot).$$

We have

$$
\begin{aligned}
L(0, e') &= \nabla(\nabla(t, t)(0, \cdot), \nabla(t, t')(0, \cdot))(e', \cdot) \\
&= \nabla(t, t')(e', \cdot)
\end{aligned}
$$

and

$$
\begin{aligned}
L(e, 0) &= \nabla(\nabla(t, t)(e, \cdot), \nabla(t, t')(e, \cdot))(0, \cdot) \\
&= \nabla(t, t')(e, \cdot)
\end{aligned}
$$

Thus, by microlinearity

$$L(e, e') = \nabla(t, t')(e + e', \cdot).$$

But then

$$
\begin{aligned}
q_{(t,e)}(p_{(t,e)}(t')) &= \nabla(\nabla(t, t)(e, \cdot), \nabla(t, t')(e, \cdot))(-e, \cdot) \\
&= L(e, -e) \\
&= \nabla(t, t')(0, .) \\
&= t'.
\end{aligned}
$$

(c) Next we show that $p_{(t,e)}(q_{t,e)}(t_1)) = t_1$. This follows from the previous point and from the fact that

$$q_{(t,e)} = p_{(\nabla(t,t)(e,\cdot), -e)}.$$

More explicitly we have :

$$
\begin{aligned}
p_{(t,e)} &(\ q_{(t,e)}(t_1)) \\
&= q_{(\nabla(t,t)(e,\cdot), -e)}(p_{(\nabla(t,t)(e,\cdot), -e)}(p_{(t,e)}(q_{(t,e)}(t_1)))) \\
&= q_{(\nabla(t,t)(e,\cdot), -e)}(q_{(t,e)}(p_{(t,e)}(q_{(t,e)}(t_1)))) \\
&= q_{(\nabla(t,t)(e,\cdot), -e)}(q_{(t,e)}(t_1)) \\
&= q_{(\nabla(t,t)(e,\cdot), -e)}(p_{(\nabla(t,t)(e,\cdot), -e)}(t_1)) \\
&= t_1.
\end{aligned}
$$

(d) Finally, the linearity of p, and thus of q, follows from its homogeneity.

∎

The notion of parallel transport described here shows the notion of vertical component of a microsquare in a different light.

Let $\gamma : D \times D \to M$ be a microsquare at m. We put, as usual, $K(\gamma) = (t_1, t_2)$. If e_1 is fixed in D, $\gamma(e_1, \cdot)$ is a tangent vector to M at $\gamma(e_1, 0) = t_1(e_1)$. Thus we can transport it, parallel to t_1, to m, which means that we consider

$$q_{(t_1, e_1)}(\gamma(e_1, \cdot)).$$

Proposition 7. *If $\gamma : D \times D \to M$ is a microsquare, we have*

$$
\begin{aligned}
q_{(t_1, e_1)}(\gamma(e_1, \cdot)) &= V(\gamma)(e_1, \cdot) \\
&= t_2 + e_1 \cdot C(\gamma).
\end{aligned}
$$

Proof: Let us consider for $(e_1, e_1') \in D(2)$

$$\nabla(\nabla(t_1, t_1)(e_1, \cdot), \gamma(e_1, \cdot))(e_1', \cdot).$$

By putting $e_1' = 0$, we get $\gamma(e_1, \cdot)$. By putting $e_1 = 0$, we get $\nabla(t_1, t_2)(e_1', \cdot)$. Thus we have

$$
\begin{aligned}
q_{(t_1, e_1)}(\gamma(e_1, \cdot)) &= \nabla(\nabla(t_1, t_1)(e_1, \cdot), \gamma(e_1, \cdot))(-e_1, \cdot) \\
&= (\gamma \,\underset{1}{\textstyle-}\, \nabla(t_1, t_2))(e_1, \cdot) \\
&= V(\gamma)(e_1, \cdot).
\end{aligned}
$$

The second equality is then included in Proposition 3.   ∎

5.3   Torsion and curvature

5.3.1   Differential forms with value in a tangent fibre bundle

We defined in Chapter 4 the differential forms with value in a (microlinear and Euclidean) R-module. We extend now this concept to a notion of differential forms with value in a tangent vector bundle.

Let M_1 and M_2 be two microlinear objects. Let φ be a fixed mapping

$$\varphi : M_1 \to M_2.$$

We are going to define a differential φ-form on M_1 with value in the tangent fibre bundle $TM_2 \xrightarrow{\pi} M_2$.

Definition 1. *A differential n-φ-form on M_1, with value in*

$$TM_2 \xrightarrow{\pi} M_2$$

is a mapping :

$$\omega : M_1^{(D^n)} \to TM_2$$

that (as in 4.1) is n-homogeneous and alternated and that makes the following diagram commutative

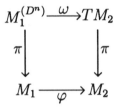

We do not immediately have the notion of an exterior differential of a n-form, because if $\tau : D^{n+1} \to M_1$ is an $(n+1)$-microcube of M_1 and marked by $(e_1, \ldots, e_{n+1}) \in D^{n+1}$, its sides are not all centered at $m = \tau(\underline{0})$. Half of the sides are centered at m, but the other half have centers varying with the e_i. The other sides thus have images through ω, that are tangent vectors to M_2 at points other than $\varphi(m)$, varying with the e_i.

Thus these different tangent vectors to M_2 cannot immediately be added up. A "transport" to the point $\varphi(m)$ would be useful. And this, as we shall see, is possible if we have a connection ∇ on M_2.

Let ω be a differential n-form on M_1 with value in the tangent fibre bundle at M_2 and let ∇ be a connection on M_2. We are going to define a differential $(n+1)$-form on M_1 again with value in a tangent fibre bundle at M_2, which will depend on the connection ∇ and that we shall call the covariant differential of ω.

Let $(\gamma, \underline{e})$ be a marked $(n+1)$-microcube. Let us recall (cf. 4.2.1) that for every i, $1 \le i \le n+1$, we consider the two marked faces

$$F_0^i(\gamma; \underline{e}) = (\gamma_0^i; (e_1, \ldots, \hat{e}_i, \ldots, e_{n+1}))$$
$$F_1^i(\gamma; \underline{e}) = (\gamma_1^i; (e_1, \ldots, \hat{e}_i, \ldots, e_{n+1}))$$

where

$$\gamma_0^i(d_1, \ldots, d_n) = \gamma(d_1, \ldots, d_{i-1}, 0, d_i, \ldots, d_n)$$
$$\gamma_1^i(d_1, \ldots, d_n) = \gamma(d_1, \ldots, d_{i-1}, e_i, d_i, \ldots, d_n).$$

We refer by $\gamma_i : D \to M_1$ to the ith edge of the $(n+1)$-microcube, that is to the tangent vector given by

$$\gamma_i(d) = \gamma(0, \ldots, 0, d, 0, \ldots, 0)$$

(where d is in i^{th} position) ; it is a tangent vector at $m = \gamma(\underline{0})$. Notice that γ_0^i is an n-microcube centered at m, whereas γ_1^i is an n-microcube centered at $\gamma_i(e_i)$. Notice that $d\varphi(\gamma_i) = \varphi \circ \gamma_i$ is a tangent vector to M_2 at $\varphi(m)$.

As $\omega(\gamma_1^i)$ is a tangent vector to M_2 at the point $\varphi(\pi(\gamma_1^i)) = \varphi(\gamma_i(e_i)) = d\varphi(\gamma_i)(e_i)$, we can consider the result from the parallel transport of this tangent vector at $\varphi(m)$ along $d\varphi(\gamma_i)$. More exactly we consider

$$q_{(d\varphi(\gamma_i), e_i)}(\omega(\gamma_1^i)) \in T_{\varphi(m)} M_2,$$

a vector that we shall denote by $\tilde{\omega}(\gamma_1^i)$. We can consider then in the R-module $T_{\varphi(m)} M_2$, the sum

$$\Phi(\gamma, \underline{e}) = \sum_{i=1}^{n+1} (-1)^i e_1 \cdot e_2, \ldots, \hat{e}_i, \ldots, e_{n+1}(\omega(\gamma_0^i) - \tilde{\omega}(\gamma_1^i)),$$

and so we can pick up again the proof of Proposition 2 of 4.2 and establish that if $\Phi(\gamma, \underline{e})$ is multilinear and alternating, there exists a unique $(n+1)$-form Ω such that

$$\Phi(\gamma, \underline{e}) = e_1 e_2 \ldots e_{n+1} \Omega(\gamma).$$

In fact multilinearity and the alternating property are satisfied by $\Phi(\gamma, \underline{e})$. To see this we proceed in the same way as in Proposition 3 of 4.2, observing that $\tilde{\omega}(\gamma_1^i)$ is a function of e_i with values in $T_{\varphi(m)}M_2$. Its value for $e_i = 0$ is $\omega(\gamma_0^i)$ and thus we can put

$$\tilde{\omega}(\gamma_1^i) = \omega(\gamma_0^i) + e_i \cdot A_i.$$

Thus we have

$$\Phi(\gamma, \underline{e}) = e_1 \cdot e_2 \cdot \ldots \cdot e_{n+1} \cdot \sum_{i=1}^{n+1} (-1)^{i+1} A_i.$$

Finally, we have the following proposition.

Proposition 1. *There is one and only one differential $(n + 1)$-form $D\omega$ on M_1 with value in the tangent fibre bundle at M_2, such that*

$$e_1 \cdot e_2 \cdot \ldots \cdot e_{n+1} D\omega(\gamma) = \sum_{i=1}^{n+1} (-1)^i e_1 \cdot e_2 \cdot \ldots \cdot \hat{e}_i \cdot \ldots \cdot e_{n+1} (\omega(\gamma^i) - \hat{\omega}(\gamma_1^i))$$

where

$$\tilde{\omega}(\gamma_1^i) = q_{(d\varphi(\gamma_i), e_i)}(\omega(\gamma_1^i)).$$

∎

Definition 2 . *The form $D\omega$ is called the exterior covariant differential of ω.*

We should note that we could have defined a notion of connection in a vector bundle and considered differential forms with value in a vector bundle equipped with a connection. In that case, the constructions of Chapter 4 would be strictly a special case of constructions so developed by associating to the R-module E the trivial vector bundle on the object M_2 reduced to a point.

5.3.2 Torsion

A first example, at first sight a trivial one, of a differential 1-form defined in M and with value in the tangnt fibre bundle $TM \to M$ is simply the identity θ :

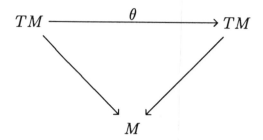

In fact, this form θ corresponds to what is classically called the "solder form" or "canonical form" (cf. [24], Chapter 3, Proposition 2.1).

Let us suppose that M is equipped with a connection ∇.

Definition 3. *The exterior covariant differential $D\theta$ of the identity form θ is called the torsion form of the connection ∇.*

Let C be the connecting mapping associated to ∇ described in 5.2.2. The fundamental property of the torsion form, which justifies its name, is that $D\theta(\gamma)$ measures, in a certain way, the difference between the microsquare γ and the symmetrical microsquare $\Sigma(\gamma)$. Of course we cannot directly compare γ and $\Sigma(\gamma)$ by strong difference because in general they don't have the same restrictions to the axes. On the other hand we can compare their image under the connecting mapping C. The result is as follows.

Proposition 2. *For every microsquare $\gamma : D \times D \to M$, we have*

$$D\theta(\gamma) = C(\gamma) - C(\Sigma(\gamma)).$$

Proof: Let us consider the marked microsquare $(\gamma; (e_1, e_2))$ and the expression

$$e_1 \cdot e_2 \cdot D\theta(\gamma)$$

which, thanks to Proposition 1, is equal to

$$e_2 \cdot (\tilde{\theta}(\gamma_1^1) - \theta(\gamma_0^1)) - e_1 \cdot (\tilde{\theta}(\gamma_1^2) - \theta(\gamma_0^2))$$

where

$$
\begin{aligned}
\gamma_0^1(d) &= \gamma(0, d) = \gamma_2(d) \\
\gamma_1^1(d) &= \gamma(e_1, d) \\
\gamma_0^2(d) &= \gamma(d, 0) = \gamma_1(d) \\
\gamma_1^2(d) &= \gamma(d, e_2).
\end{aligned}
$$

We have $\theta(\gamma_0^1) = \gamma_0^1 = \gamma_2$ and by virtue of Proposition 8 of 5.2,

$$
\begin{aligned}
\tilde{\theta}(\gamma_1^1) &= q_{(\gamma_1, e_1)}(\gamma(e_1, \cdot)) \\
&= V(\gamma)(e_1, \cdot) \\
&= \gamma_2 + e_1 \cdot C(\gamma).
\end{aligned}
$$

Thus the first term of $e_1 \cdot e_2 \cdot D\theta(\gamma)$ is

$$
\begin{aligned}
e_2(\tilde{\theta}(\gamma_1^1) - \theta(\gamma_0^1)) &= e_2 \cdot (\gamma_2 + e_1 \cdot C(\gamma) - \gamma_2) \\
&= e_1 \cdot e_2 \cdot C(\gamma).
\end{aligned}
$$

In the same way, by symmetry, the second term is

$$
\begin{aligned}
-e_1 \cdot (\tilde{\theta}(\gamma_1^2) - \theta(\gamma_0^2)) &= -e_1(\gamma_1 + e_2 \cdot C(\Sigma(\gamma)) - \gamma_1) \\
&= -e_1 \cdot e_2 C(\Sigma(\gamma)).
\end{aligned}
$$

Thus we have for every e_1 and e_2 of D

$$
e_1 \cdot e_2 \cdot D\theta(\gamma) = e_1 \cdot_2 (C(\gamma) - C(\Sigma(\gamma))),
$$

hence the thesis. ∎

The notion of torsion forms, expressed in terms of vector fields, leads to the notion of torsion tensor fields.

Definition 4. *We call* torsion, *or* torsion tensor field, *the mapping*

$$
T : \mathcal{X}(M) \times \mathcal{X}(M) \to \mathcal{X}(M)
$$

described by

$$
T(X, Y)(m) = D\theta((Y \star X)_m)
$$

for the vector fields X and Y.

We have the convenient torsion expression.

Proposition 3. *For vector fields X and Y we have :*

$$
T(X, Y) = \nabla_X Y - \nabla_Y X - [X, Y].
$$

Proof: Taking into account Proposition 2 and the expression of the connecting mapping C given in Proposition 4 of 5.2, it reduces to a simple computation in an affine space. We have in fact

$$
\begin{aligned}
T(X,Y) &= C(Y \star X) - C(\Sigma(Y \star X)) \\
&= (Y \star X \dot{-} \nabla(X,Y)) - (\Sigma(Y \star X) \dot{-} \nabla(Y,X)).
\end{aligned}
$$

By Proposition 6 of 5.2, we have

$$
\begin{aligned}
\nabla_X Y - \nabla_Y X &= C(Y \star X) - C(X \star Y) \\
&= (Y \star X \dot{-} \nabla(X,Y)) - (X \star Y \dot{-} \nabla(Y,X)).
\end{aligned}
$$

Thus we have

$$
\begin{aligned}
\nabla_X Y - \nabla_Y X - T(X,Y) &= (\Sigma(Y \star X) \dot{-} \nabla(Y,X)) \\
&\quad -(X \star Y \dot{-} \nabla(Y,X)) \\
&= \Sigma(Y \star X) \dot{-} X \star Y,
\end{aligned}
$$

which, thanks to (3.4, Prop. 8), equals $[X,Y]$ as we had to show.   ∎

The same type of trivial computation in an affine space also yields the following expression of the torsion form and the torsion.

Proposition 4. *The torsion is given by*

$$
T(X,Y) = \Sigma(\nabla(Y,X) \dot{-} \nabla(X,Y)
$$

and the torsion form by

$$
D\theta(\gamma) = \Sigma \nabla K \Sigma \gamma \dot{-} \nabla K \gamma.
$$

Proof: We have

$$
\begin{aligned}
D\theta(\gamma) &= C(\gamma) - C(\Sigma(\gamma)) \\
&= (\gamma \dot{-} \nabla K \gamma) - (\Sigma \gamma \dot{-} \nabla K \Sigma \gamma)
\end{aligned}
$$

which, by (3.4, Proposition 6), is equal to

$$
(\gamma \dot{-} \nabla K \gamma) - (\gamma \dot{-} \Sigma \nabla K \Sigma \gamma)
$$

and thus $\Sigma \nabla K \Sigma \gamma \dot{-} \nabla K \gamma$ as announced. The claim about the torsion is a particular case.   ∎

We have the following corollaries.

Corollary 5. *A connection is symmetrical if and only if its torsion is null and if and only if its torsion form is null.* ∎

The second corollary is the "tensor" character of the torsion T, that is its bilinearity.

Corollary 6. *The application T is R^M-bilinear.*

Proof: Thanks to Proposition 4, we have

$$
\begin{aligned}
T(f \cdot X, Y) &= \Sigma \nabla(Y, f \cdot X) \dot{-} \nabla(F \cdot X, Y) \\
&= \Sigma(f \underset{2}{\cdot} \nabla(Y, X)) \dot{-} f \underset{1}{\cdot} \nabla(X, Y) \\
&= f \underset{1}{\cdot} \Sigma \nabla(Y, X) \dot{-} f \underset{1}{\cdot} \nabla(X, Y) \\
&= f \cdot T(X, Y),
\end{aligned}
$$

and in the same way at Y. ∎

5.3.3 Curvature

The connecting mapping

$$ C : M^{D \times D} \to M^D $$

defined by

$$ C(\gamma) = \gamma \dot{-} \nabla K \gamma $$

(cfr. 5.2.2) can be looked upon as a differential 1-form, defined on the manifold M^D and with value in the tangent fibre bundle M^D. It is a 1-π-form where $\pi : M^D \to M$ is the projection, because the diagram

$$
\begin{array}{ccc}
(M^D)^D \simeq M^{D \times D} & \xrightarrow{\;\;C\;\;} & M^D \\
{\scriptstyle M^{i_2}}\downarrow & & \downarrow{\scriptstyle \pi} \\
M^D & \xrightarrow[\;\;\pi\;\;]{} & M
\end{array}
$$

commutes and C is R-linear (cfr. 5.2, Proposition 5). That is the reason why the connecting mapping C has also been called "connection form" (cfr. Definition in 5.2.2).

We shall look now at the covariant exterior differential of the 1-degree π-form, or connection form C.

Definition 5. *We call the π-form of degree 2*

$$\Omega = DC.$$

the curvature form.

The curvature form Ω is a π-form of degree 2, defined on the manifold M^D and with value in the tangent fibre bundle at M :

$$(M^D)^{D\times D} \xrightarrow{\ \Omega\ } M^D$$

$$\downarrow \qquad\qquad \downarrow \pi$$

$$M^D \xrightarrow[\ \pi\ ]{} M$$

We shall first indicate an explicit computation of Ω :

Proposition 7. *The curvature form is given by*

$$\Omega = C \circ C^D - C \circ C^D \circ \Sigma.$$

Proof: Let $(\tau; (e_1, e_2))$ be a microsquare in M^D. Let us consider the expression

$$e_1 \cdot e_2 \cdot \Omega(\tau).$$

By virtue of Proposition 1, it is equal to

$$e_2 \cdot (\tilde{C}(\tau_1^1) - C(\tau_0^1)) - e_1 \cdot (\tilde{C}(\tau_1^2) - C(\tau_0^2))$$

where

$$
\begin{aligned}
\tau_0^1(d_2, d_3) &= \tau(0, d_2)(d_3) \\
\tau_1^1(d_2, d_3) &= \tau(e_1, d_2)(d_3) \\
\tau_0^2(d_1, d_3) &= \tau(d_1, 0)(d_3) \\
\tau_1^2(d_1, d_3) &= \tau(d_1, e_2)(d_3).
\end{aligned}
$$

Let us describe $\tilde{C}(\tau_1^1)$. As was seen in 5.3.1, we have

$$\tilde{C}(\tau_1^1) = q_{(d\pi(\tau_1),e_1)}(C(\tau_1^1))$$

where τ_1 is the first axis of τ, in other words the tangent vector to M^D given by

$$\tau_1(d_1) = \tau(d_1,0)(\cdot).$$

As $\pi : M^D \to M$ associates to a tangent vector its contact point, $d\pi(\tau_1)$ is the tangent vector to M given by

$$d\pi(\tau_1)(d_1) = \tau(d_1,0)(0).$$

Let us consider all the microsquares

$$\gamma = C^D(\tau)$$

given by

$$\gamma(d_1,d) = C(\tau(d_1,\cdot)(\cdot))(d).$$

We apply Proposition 7 of 5.2. The axes of γ are given by

$$
\begin{aligned}
\gamma_1(e_1) &= \gamma(e_1,0) = C(\tau(e_1,\cdot)(\cdot))(0) \\
&= C(\tau_1^1)(0) \\
&= \tau(e_1,0)(0) \\
&= d\pi(\tau_1)(e_1)
\end{aligned}
$$

and

$$
\begin{aligned}
\gamma_2(d) = \gamma(0,d) &= C(\tau(0,\cdot)(\cdot))(d) \\
&= C(\tau_0^1)(d),
\end{aligned}
$$

thus $\gamma_1 = d\pi(\tau_1)$ and $\gamma_2 = C(\tau_0^1)$. Proposition 7 of 5.2 tells us that

$$q_{(\gamma_1,e_1)}(\gamma(e_1,\cdot)) = \gamma_2 + e_1 \cdot C(\gamma).$$

As $\gamma(e_1,\cdot) = C(\tau(e_1,\cdot)(\cdot))(\cdot) = C(\tau_1^1)$, we get

$$
\begin{aligned}
\tilde{C}(\tau_1^1) &= q_{(d\pi(\tau_1),e_1)}(C(\tau_1^1)) \\
&= q_{(\gamma_1,e_1)}(\gamma(e_1,\cdot)) \\
&= \gamma_2 + e_1 \cdot C(\gamma) \\
&= C(\tau_0^1) + e_1 \cdot C(\gamma).
\end{aligned}
$$

Thus we have

$$\tilde{C}(\tau_1^1) - C(\tau_0^1) = e_1 \cdot C(\gamma)$$
$$= e_1 \cdot C(C^D(\tau)).$$

Thanks to symmetry we have in the same way

$$\tilde{C}(\tau_1^2) - C(\tau_0^2) = e_2 \cdot C(C^D(\Sigma\tau))$$

where

$$(\Sigma\tau)(d_2, d_1)(d_3) = \tau(d_1, d_2)(d_3).$$

Thus we have

$$e_1 \cdot e_2 \cdot \Omega(\tau) = e_1 \cdot e_2 \cdot (C(C^D(\tau)) - C(C^D(\Sigma\tau))),$$

hence the thesis. ∎

The notion of curvature form, expressed in terms of vector fields, leads us to the notion of curvature tensor fields.

Definition 6. *We call curvature, or curvature tensor field the mapping*

$$R : \mathcal{X}(M) \times \mathcal{X}(M) \times \mathcal{X}(M) \to \mathcal{X}(M)$$

described by

$$R(X, Y, Z)(m) = \Omega((Z \star Y \star X)_m)$$

where

$$(Z \star Y \star X)_m(d_1, d_2)(d_3) = (Z_{d_3} \circ Y_{d_2} \circ X_{d_1})(m).$$

We can express the curvature algebraically as the extend to which the covariant derivative for the Lie bracket fails to be homomorphic

Proposition 8. *For all vector fields X, Y, and Z, we have :*

$$R(X, Y, Z) = \nabla_X(\nabla_Y Z) - \nabla_Y(\nabla_X Z) - \nabla_{[X,Y]}(Z).$$

Proof: Let us compute the different terms of the formula to be proved.

(a)

$$R(X, Y, Z)_m = \Omega((Z \star Y \star X)_m)$$
$$= (C \circ C^D - C \circ C^D \circ \Sigma)((Z \star Y \star X)_m)$$

by virtue of Proposition 7.

We use again the convention about *tilde* as given at the beginning of Section 4.5.

Thus we can write $(C \circ C^D)((Z \star Y \star X)_m)$ in the form of

$$C[C((Z_{\underline{d_3}} \circ Y_{\underline{d_2}} \circ X_{\underline{d_1}})(m))(\underline{d})]. \tag{1}$$

In the same way we write $(C \circ C^D \circ \Sigma)((Z \star Y \star X)_m)$ in the form

$$C[C((Z_{\underline{d_3}} \circ Y_{\underline{d_2}} \circ X_{\underline{d_1}})(m))(\underline{d})]. \tag{2}$$

(b) According to Proposition 6 of 5.2, we have

$$\begin{aligned}
\nabla_X(\nabla_Y Z) &= C[((\nabla_Y Z)_{\underline{d}} \circ X_{\underline{d_1}})(m)] \\
&= C[C((Z_{\underline{d_3}} \circ Y_{\underline{d_2}} \circ X_{\underline{d_1}})(m))(\underline{d})]
\end{aligned} \tag{3}$$

and in the same way

$$\begin{aligned}
\nabla_Y(\nabla_X Z) &= C[((\nabla_X Z)_{\underline{d}} \circ Y_{\underline{d_2}})(m)] \\
&= C[C((Z_{\underline{d_3}} \circ X_{\underline{d_1}} \circ Y_{\underline{d_2}})(m)(\underline{d})].
\end{aligned} \tag{4}$$

Finally

$$\nabla_{[X,Y]} Z = C((Z_{\underline{d_3}} \circ [X,Y]_{\underline{d}})(m)). \tag{5}$$

We must show that

$$(1) - (2) = (3) - (4) - (5)$$

or, as $(1) = (3)$, that

$$(2) - (4) = (5).$$

We begin with the computation of $(2) - (4)$

$$C[C((Z_{\underline{d_3}} \circ Y_{\underline{d_2}} \circ X_{\underline{d_1}}(m))(\underline{d})] - C[C((Z_{\underline{d_3}} \circ X_{\underline{d_1}} \circ Y_{\underline{d_2}})(m)(\underline{d})].$$

Put

$$\begin{aligned}
\delta_1(d_2, d) &= C((Z_{\underline{d_3}} \circ Y_{d_2} \circ X_{\underline{d_1}})(m))(d) \\
\delta_2(d_2, d) &= C((Z_{\underline{d_3}} \circ X_{\underline{d_1}} \circ Y_{d_2})(m))(d).
\end{aligned}$$

We have

$$
\begin{aligned}
\delta_1(d_2, 0) &= C((Z_{\underline{d}_3} \circ Y_{d_2} \circ X_{\underline{d}_1})(m))(0) \\
&= Y_{d_2}(m) \\
\delta_2(d_2, 0) &= C((Z_{\underline{d}_3} \circ X_{\underline{d}_1} \circ Y_{d_2})(m))(0) \\
&= Y_{d_2}(m).
\end{aligned}
$$

As C is R-linear for the vector fibre bundle structure $(\underset{i}{\,\cdot\,} , \underset{1}{+})$, of $M^{D \times D}$ over M^D (5.2, Proposition 5), we have

$$
\begin{aligned}
(2) - (4) &= C(\delta_1) - C(\delta_2) \\
&= C(\delta_1 \underset{1}{-} \delta_2).
\end{aligned}
$$

Let us describe $(\delta_1 \underset{1}{-} \delta_2)(d_2, d)$. Put

$$
\varphi(d_2, d_2', d) = C((Z_{\underline{d}_3} \circ Y_{d_2} \circ X_{\underline{d}_1} \circ Y_{-d_2'}(m))(d)
$$

for $(d_2, d_2') \in D(2)$. For $d_2' = 0$, we have

$$
\varphi(d_2, 0, d) = \delta_1(d_2, d)
$$

and, for $d_2 = 0$, we have

$$
\varphi(0, d_2', d) = \delta_2(-d_2', d)
$$

and thus

$$
\begin{aligned}
(\delta_1 \underset{1}{-} \delta_2)(d_2, d) &= \varphi(d_2, d_2, d) \\
&= C((Z_{\underline{d}_3} \circ Y_{d_2} \circ X_{\underline{d}_1} \circ Y_{-d_2})(m))(d).
\end{aligned}
$$

This microsquare $\delta = \delta_1 \underset{1}{-} \delta_2$ is, by construction, vertical :

$$
\begin{aligned}
\delta(d_2, 0) &= C((Z_{\underline{d}_3} \circ Y_{d_2} \circ X_{\underline{d}_1} \circ Y_{-d_2}(m))(0) \\
&= (Y_{d_2} \circ Y_{-d_2})(m) \\
&= m.
\end{aligned}
$$

The axis of δ is

$$
\delta(0, d) = C((Z_{\underline{d}_3} \circ X_{\underline{d}_1})(m))(d)
$$

and its deviation is $C(\delta)$ ($= (2) - (4)$). Thus we have

$$\delta(d_2, d) = (C((Z_{\underline{d_3}} \circ X_{\underline{d_1}})(m)) + d_2 \cdot C(\delta))(d)$$

or

$$
\begin{aligned}
d_2 \cdot C(\delta) &= \delta(d_2, \underline{d}) - C((Z_{\underline{d_3}} \circ X_{\underline{d_1}})(m))(\underline{d}) \\
&= C((Z_{\underline{d_3}} \circ Y_{d_2} \circ X_{\underline{d_1}} \circ Y_{-d_2})(m)) - C((Z_{\underline{d_3}} \circ X_{\underline{d_1}})(m)) \\
&= C(\epsilon_1) - C(\epsilon_2)
\end{aligned}
$$

with

$$
\begin{aligned}
\epsilon_1(d_1, d_3) &= (Z_{d_3} \circ Y_{d_2} \circ X_{d_1} \circ Y_{-d_2})(m) \\
\epsilon_2(d_1, d_3) &= (Z_{d_3} \circ X_{d_1})(m).
\end{aligned}
$$

As

$$\epsilon_1(0, d_3) = Z_{d_3}(m) = \epsilon_2(0, d_3),$$

we can once again use the R-linearity of C for the structure $(\underset{1}{,} \underset{1}{+})$ and get

$$d_2 \cdot C(\delta) = C(\epsilon_1 \underset{1}{-} \epsilon_2).$$

We compute explicitly $\epsilon_1 \underset{1}{-} \epsilon_2$ by putting

$$\psi(d_1, d_1', d_3) = (Z_{d_3} \circ X_{-d_1'} \circ Y_{d_2} \circ X_{d_1} \circ Y_{-d_2})(m)$$

for $(d_1, d_1') \in D(2)$. We find

$$
\begin{aligned}
\psi(d_1, 0, d_3) &= (Z_{d_3} \circ Y_{d_2} \circ X_{d_1} \circ Y_{-d_2})(m) = \epsilon_1(d_1, d_3) \\
\psi(0, d_1', d_3) &= (Z_{d_3} \circ X_{-d_1'})(m) = \epsilon_2(-d_1', d_3)
\end{aligned}
$$

and thus

$$
\begin{aligned}
(\epsilon_1 \underset{1}{-} \epsilon_2)(d_1, d_3) &= \psi(d_1, d_1, d_3) \\
&= (Z_{d_3} \circ X_{-d_1} \circ Y_{d_2} \circ X_{d_1} \circ Y_{-d_2})(m) \\
&= (Z_{d_3} \circ [X, Y]_{d_1 \cdot d_2})(m).
\end{aligned}
$$

Thus we have

$$d_2 \cdot C(\delta) = C((Z_{\underline{d_3}} \circ [X, Y]_{\underline{d_1} \cdot d_2})(m)).$$

But C is ($\underset{i}{\cdot}$)-homogeneous and thus

$$d_2 \cdot C(\delta) = d_2 \cdot C((Z_{\underline{d}_3} \circ [X,Y]_{\underline{d}_1})(m)).$$

Thus we do have

$$(2) - (4) = C(\delta) = C((Z_{\underline{d}_3} \circ [X,Y]_{\underline{d}_1})(m) = (5)$$

and the theorem is proved.   ∎

5.4   Commented bibliography

1) The first to approach connection theory in synthetic differential geometry are Kock and Reyes in [36](1979). The punctual connections of 5.1.1, are an adaptation of that work. The same [36] also contains steps in the direction of a theory of covariant derivatives but the Koszul law is established only under an additional hypothesis of local triviality. A proof of the Koszul law was made by Kock and Lavendhomme in [35](1984) (thanks to the affine structure of microsquares).

2) The link with the notion of spray is tackled by Bunge in collaboration with Sawyer in [10](1984). Under certain hypotheses, the authors establish there the theorem of Ambrose-Palais-Singer on the bijection between symmetrical connections and sprays. The presentation and hypotheses were simplified in [26](1983) by A. Kock, and in [35](1984) by Kock and Lavendhomme.

The theory of connections is picked up and developed by Moerdijk and Reyes in [52] (1991) (with parts of this work appearing in preprint form in 1985).

The notion of connecting mapping is already used in [36] and recurs in [10], [35] and [52].

3)In [36], under rather strong local triviality hypotheses, the curvature and torsion tensors are introduced. The situation for torsion is clarified in [35]. The notion of differential form with value in a vector fibre bundle equipped with a connection, as well as the essential of the results in Section 5.3, were in [41](1987).

Let us point that a combinatorial theory of connections is developed by Kock [34](1984).

Chapter 6

Global actions

6.1 Lie objects

It appeared in the thread of preceding chapters that objects defined in a pointwise manner give rise to some global actions. Such was the case of vector fields, differential forms or connections. We will give here a more systematic account of these situations.

6.1.1 Definition of Lie objects

In the pointwise manner, a vector field X on a microlinear object M, associates with any point m in M, a tangent vector X_m at this point m (cf. 3.2.1). The corresponding global action view is that of a derivation $D : R^M \to R^M$. This is a more general gadget as we saw that the Lie operator

$$L : \mathcal{X}(M) \to \mathrm{Der}(R^M)$$

is a Lie homomorphism but not, in general, an isomorphism (cf. 3.3.2).

The global view may be seen as algebraically more funny. For example, that $\mathrm{Der}(R^M)$ is a Lie algebra is trivial by the very definition :

$$[D_1, D_2] = D_1 \circ D_2 - D_2 \circ D_1.$$

In general, the pointwise method is more appropriate for the simulation of classical differential geometry. Note however that the cases where the two coincide are important enough to justify the study of the global view.

We describe here a common algebraization of the two situations.

Here R is just a unitary commutative ring which is a $\mathbb{Q}$-algebra. Of course we may also accept $(K - L)$ or $(K - W)$-axiom.

Definition 1. *A Lie object (A, L) is composed of*

- *a unitary, associative and commutative R-algebra A ;*

- *a Lie R-algebra L ;*

- *a structure of A-module on L denoted*

$$A \times L \to L : (a, X) \to a \cdot X ;$$

- *a R-linear action of L on A, denoted*

$$L \times A \to A : (X, a) \to X(a).$$

We postulate three specific axioms.

(1) *Lie-module axiom* :

$$\forall X, Y \in L, \forall a \in A : [X, Y](a) = X(Y(a)) - Y(X(a)).$$

(2) *Derivation axiom* :

$$\forall X \in L, \forall a, b \in A \quad X(ab) = X(a)b + aX(b).$$

(3) *Koszul's axiom* :

$$\forall X, Y \in L, \forall a \in A \quad [X, a \cdot Y] = a \cdot [X, Y] + X(a) \cdot Y.$$

Examples.

1) Let us first consider the general algebraic situation. Let A be a unitary, associative and commutative algebra and $\mathrm{Der}(A)$ the Lie algebra of derivations of A. For evident actions, $(A, \mathrm{Der}\, A)$ is a Lie object.

2) In the global view on synthetic differential geometry, we can take for R the real line, for A the algebra R^M of real valued functions on M. $(R^M, \mathrm{Der}(R^M))$ is a Lie object, the standard global one on M.

3) In the pointwise view, we take also $A = R^M$, but we take $L = \mathcal{X}(M)$. $(R^M, \mathcal{X}(M))$ is a Lie object, the standard pointwise one on M.

6.1.2 Differential forms

A differential n-form, as we defined in 4.1.1, acts in a pointwise manner on n-microcubes. We would be satisfied if only we had a global action on fields of n-microcubes.

A field of n-microcubes, or a n-dimensional distribution, on M is a map

$$\Gamma : M \to M^{(D^n)}$$

such that $\Gamma(m)(\underline{o}) = m$ for all m in M. For $n = 1$ we find again a vector field. Let us denote by $\mathcal{X}^{(n)}(M)$, the set of fields of n-microcubes. The maps

$$M^{i_k} : M^{(D^n)} \to M^{(D^{n-1})}$$

induce by composition maps

$$\mu_k : \mathcal{X}^{(n)}(M) \to \mathcal{X}^{(n-1)}(M).$$

We have thus vector bundles whose fibres are R^M-modules. If f is in R^M and Γ in $\mathcal{X}^{(n)}(M)$, we have

$$(f \underset{k}{\cdot} \Gamma)(x)(d_1, \ldots, d_n) = \Gamma(x)(d_1, \ldots, f(x) \cdot d_k, \ldots, d_n).$$

If σ is a permutation of $\{1, 2, \ldots, n\}$, we also define

$$\Sigma : \mathcal{X}^{(n)}(M) \to \mathcal{X}^{(n)}(M)$$

by permutation of the axes, $\Sigma(\Gamma)(x) = \Sigma(\Gamma(x))$.

We define now global differential n-forms :

Definition 2. *A singular global differential n-form on M with values in a R^M-module E is a map*

$$\omega : \mathcal{X}^{(n)}(M) \to E$$

such that

a) *ω is n-homogeneous : $\omega(f \underset{k}{\cdot} \Gamma) = f \cdot \omega(\Gamma)$ for every k, $(1 \le k \le n)$, every Γ in $\mathcal{X}^{(n)}(M)$ and every f in R^M.*

b) *ω is alternated : $\omega(\Sigma(\Gamma)) = \epsilon_\sigma \omega(\Gamma)$ for every Γ in $\mathcal{X}^{(n)}(M)$.*

Proposition 1. *The differential n-form on M with values in a R-module E induce global differential n-forms on M with values in E^M.*

Proof: If ω is pointwise defined, we merely set $\omega(\Gamma)(x) = \omega(\Gamma(x))$ for every field of n-microcubes Γ. ∎

The classical differential n-forms operate on n-uples of tangent vectors (cf. 4.1.2). The global point of view is thus given by the following definition.

Definition 3. *A classical global differential n-form on M with values in a R^M-module E is a map*

$$\omega : \mathcal{X}(M) \times \ldots \times \mathcal{X}(M) \to E$$

which is n-homogeneous and alternated.

We have trivially the following

Proposition 2. *A classical (and thus also singular) differential n-form on M with values in a R-module E gives a classical global differential n-form on M whith values in E^M.* ∎

Actually in the pointwise case, the choice of singular n-forms rather that classical ones, was justified because the given definition of exterior differential does not work for classical forms but only for singular forms. But from the global point of view, it is not the case. We can define an exterior differential of a classical global differential n-form with values in R^M.

Definition 4. *Let ω be a (classical) global differential n-form on M with values in R^M. The exterior differential of ω is given by*

$$d\omega(X_0, X_1, \ldots, X_n) = \sum_{i=0}^{n}(-1)^i X_i(\omega(X_0, \ldots, \hat{X}_i, \ldots, X_n))$$

$$+ \sum_{0 \le i < j \le n}(-1)^{i+j}\omega([X_i, X_j], \ldots, \hat{X}_i, \ldots, \hat{X}_j, \ldots, X_n)$$

for $X_0, \ldots, X_n$ in $\mathcal{X}(M)$.

A classical calculation shows that $d\omega$ is a $(n+1)$-form. Definition 4 is known as Koszul's formula for the exterior differential. We must show, of course, that this agrees with the exterior differential of singular n-forms if ω arises from such a form. Let G be the map sending a singular differential form to a global form. We write

$$G : \Omega(M) \to \Omega_g(M).$$

Proposition 3. *The map G is compatible with the exterior differentials : $G(d\omega) = d(G(\omega))$. In other words, G is a morphism of complexes, the diagrams*

$$
\begin{array}{ccc}
\Omega^n(M) & \xrightarrow{\ G^n\ } & \Omega_g^n(M) \\
\downarrow{\scriptstyle d} & & \downarrow{\scriptstyle d} \\
\Omega^{n+1}(M) & \xrightarrow{\ G^{n+1}\ } & \Omega_g^{n+1}(M)
\end{array}
$$

commute.

Proof: We give only explicit proof for $n = 1$. It is a good exercise to write a proof for $n = 2$ or for any n. We must see that

$$d\omega(X_0, X_1) = X_0(\omega(X_1)) - X_1(\omega(X_0)) - \omega([X_0, X_1]).$$

Let $e \in D$, we have

$$e \cdot d\omega(X_0, X_1) = e \cdot DF^0(0) - e \cdot DF^1(0)$$

(see 4.2.3, proposition 4) where

$$\begin{aligned} F^0(e) &= \omega(X_{1,d} \circ X_{0,e}) \\ F^1(e) &= \omega(X_{1,e} \circ X_{0,d}) \end{aligned}$$

and thus

$$\begin{aligned} e \cdot DF^0(0) &= \omega(X_{1,d} \circ X_{0,e}) - \omega(X_1) \\ e \cdot DF^1(0) &= \omega(X_{1,e} \circ X_{0,d}) - \omega(X_0). \end{aligned}$$

We have also

$$\begin{aligned} e \cdot X_0(\omega(X_1)) &= \omega(X_{1,d} \circ X_{0,e}) - \omega(X_1) \\ e \cdot X_1(\omega(X_0)) &= \omega(X_{0,d} \circ X_{1,e}) - \omega(X_0). \end{aligned}$$

The thesis reduces, for every x in M, to

$$\omega(X_{0,d} \circ X_{1,e})(x) - \omega(X_{1,e} \circ X_{0,d})(x) = -e \cdot \omega([X_0, X_1])(x).$$

We consider, for e in D and (d, d') in $D(2)$,

$$(X_{0,d} \circ X_{1,e} \circ [X_0, X_1]_{e \cdot d})(x).$$

For $d' = 0$ this is $X_{0,d}(X_{1,e}(x))$ and for $d = 0$ we have $X_{1,e}([X_0, X_1]_{e \cdot d'}(x))$. The vectors $X_{0,d}(X_{1,e}(x))$ and $X_{1,e}([X_0, X_1]_{e \cdot d}(x))$ are tangent at $X_{1,e}(x)$. Their sum is thus given, for d in D, by

$$(X_{0,d} \circ X_{1,e} \circ [X_0, X_1]_{e \cdot d})(x)$$

which is $(X_{1,e} \circ X_{0,d})(x)$.

But ω is linear on $T_{X_{1,e}(x)}$ and we have

$$\omega(X_{1,e} \circ X_{0,d})(x) = \omega(X_{0,d} \circ X_{1,e})(x) + \omega(X_{1,e} \circ [X_0, X_1]_{e \cdot d})(x).$$

Finally observe that

$$\begin{aligned} \omega(X_{1,e} \circ [X_0, X_1]_{e \cdot d}) &= e \cdot \omega(X_{1,e} \circ [X_0, X_1]_d) \\ &= e \cdot \omega([X_0, X_1]_d) \end{aligned}$$

and the theorem is proved.   ∎

The result just proved leads us to consider differential forms in the context of Lie objects.

Definition 5. *Let (A, L) be a Lie object and E an A-module. A p-form on (A, L) with values in E is a map*

$$\omega : L^p \to E$$

which is A-multilinear and antisymmetrical.

We denote by $\Omega_A^p(L, E)$ the A-module of p-forms with values in E. If E is A, we write $\Omega_A^p(L)$ and we say these p-forms are scalar.

Proposition 4. $\Omega_A(L)$, *the direct sum of the* $\Omega_A^p(L)$, *is a graduated algebra for the exterior product defined by*

$$(\omega \wedge \omega')(X_1, \ldots, X_{p+q})$$
$$= \frac{1}{p!q!} \sum_{\sigma \in S_{p+q}} \epsilon_\sigma \omega(X_{\sigma(1)}, \ldots, X_{\sigma(p)}) \omega'(X_{\sigma(p+1)}, \ldots, X_{\sigma(p+q)})$$

for $\omega \in \Omega_A^p(L)$ and $\omega' \in \Omega_A^q(L)$. Moreover, if E is a A-module, litterally the same formula provide $\Omega_A(L, E)$ with a structure of graduated $\Omega_A(L)$-module.

Proof: A calculation with permutations. ∎

The following proposition is also classical :

Proposition 5. *The graduated algebra $\Omega_A(L)$ is a differential graded algebra, the exterior differential being defined by*

$$(d\omega)(X_0, \ldots, X_p) = \sum_{i=0}^{p} (-1)^i X_i(\omega(X_0, \ldots, \hat{X}_i, \ldots, X_p))$$
$$+ \sum_{0 \le i < j \le p} (-1)^{i+j} \omega([X_i, X_j], \ldots, \hat{X}_i, \ldots, \hat{X}_j, \ldots, X_p)$$

(the formula of definition 4).

Proof: Consider only the case $p = 1$.

We must show $d(d\omega)(X_0, X_1, X_2) = 0$. We have

$$
\begin{aligned}
d(d\omega)(X_0, X_1, X_2) &= X_0(d\omega(X_1, X_2)) - X_1(d\omega(X_0, X_2)) \\
&\quad + X_2(d\omega(X_0, X_1)) - d\omega([X_0, X_1], X_2) \\
&\quad + d\omega([X_0, X_2], X_1) - d\omega([X_1, X_2], X_0) \\
&= X_0\big(X_1(\omega(X_2))\big) - X_0\big(X_2(\omega(X_1))\big) - X_0(\omega([X_1, X_2])) \\
&\quad - X_1\big(X_0(\omega(X_2))\big) + X_1\big(X_2(\omega(X_0))\big) + X_1(\omega([X_0, X_2])) \\
&\quad + X_2\big(X_0(\omega(X_1))\big) - X_2\big(X_1(\omega(X_0))\big) - X_2(\omega([X_0, X_1])) \\
&\quad - [X_0, X_1](\omega(X_2)) + X_2(\omega([X_0, X_1])) \\
&\quad + [X_0, X_2](\omega(X_1)) - X_1(\omega([X_0, X_2])) \\
&\quad - [X_1, X_2](\omega(X_0)) + X_0(\omega([X_1, X_2]))
\end{aligned}
$$

and this is zero.   ∎

Definition 6. *If $X \in L$ and $\omega \in \Omega_A^{p+1}(L, E)$, the interior product $i_X(\omega)$ is the p-form given by*

$$
i_X(\omega)(X_1, \ldots, X_p) = \omega(X, X_1, \ldots, X_p).
$$

We can also put $i_X(v) = 0$ for $v \in E$.

Definition 7. *If $X \in L$ and $\omega \in \Omega_A^p(L)$, the Lie derivative of ω, $\mathcal{L}_X(\omega)$ is the p-form given by*

$$
\begin{aligned}
\mathcal{L}_X(\omega)(X_1, \ldots, X_p) &= X(\omega(X_1, \ldots, X_p)) \\
&\quad - \sum_{i=1}^{p} \omega(X_1, \ldots, [X, X_i], \ldots, X_p).
\end{aligned}
$$

The following proposition is a synthesis of what is called the H. Cartan calculus.

Proposition 6.

1) d is an antiderivation of degree $+1$.

2) i_X is an antiderivation of degree -1.

3) $\mathcal{L}_X$ is a derivation of degree 0.

4) $\mathcal{L}_X \circ d = d \circ \mathcal{L}_X$.

5) $\mathcal{L}_X = i_X \circ d + d \circ i_X$.

6) $[\mathcal{L}_X, \mathcal{L}_Y] = \mathcal{L}_{[X,Y]}$.

7) $[\mathcal{L}_X, i_Y] = i_{[X,Y]}$.

Proof: Proofs are exclusively computational. To avoid tedious notations, we give it only for 1-forms, but the generalization to n-forms is straightforward.

1) For $\omega, \omega' \in \Omega_A^1(L)$ we have

$$d(\omega \wedge \omega') = d\omega \wedge \omega' - \omega \wedge d\omega'.$$

The lefthand side is, in fact, given by

$$
\begin{aligned}
d(\omega \wedge \omega')(X_0, X_1, X_2) &= X_0(\omega(X_1)\omega'(X_2)) \\
&- X_0(\omega(X_2)\omega'(X_1)) - X_1(\omega(X_0)\omega'(X_2)) \\
&+ X_1(\omega(X_2)\omega'(X_0) + X_2(\omega(X_0)\omega'(X_1)) \\
&- X_2(\omega(X_1)\omega'(X_0)) - \omega([X_0, X_1])\omega'(X_2) \\
&+ \omega(X_2)\omega'([X_0, X_1]) + \omega([X_0, X_2])\omega'(X_1) \\
&- \omega(X_1)\omega'([X_0, X_2]) - \omega([X_1, X_2])\omega'(X_0) \\
&+ \omega(X_0)\omega'([X_1, X_2]) \\
&= (X_0(\omega(X_1)) - X_1(\omega(X_0)) - \omega([X_0, X_1])) \cdot \omega'(X_2) \\
&+ (X_2(\omega(X_0)) - X_0(\omega(X_2)) - \omega([X_2, X_0])) \cdot \omega'(X_1) \\
&+ (X_1(\omega(X_2)) - X_2(\omega(X_1)) - \omega([X_1, X_2])) \cdot \omega'(X_0) \\
&+ \omega(X_0) \cdot (X_1(\omega'(X_2)) - X_2(\omega'(X_1)) - \omega'([X_1, X_2])) \\
&+ \omega(X_2) \cdot (X_0(\omega'(X_1)) - X_1(\omega'(X_0)) - \omega'([X_0, X_1])) \\
&+ \omega(X_1) \cdot (X_2(\omega'(X_0)) - X_0(\omega'(X_2)) - \omega'([X_2, X_0])) \\
&= (d\omega \wedge \omega')(X_0, X_1, X_2) - (\omega \wedge d\omega')(X_0, X_1, X_2).
\end{aligned}
$$

2) For ω and ω' of degree 1, we have

$$
\begin{aligned}
i_X(\omega \wedge \omega')(Y) &= \omega(X)\omega'(Y) - \omega(Y)\omega'(X) \\
&= (i_X(\omega) \wedge \omega')(Y) - (\omega \wedge i_X\omega')(Y).
\end{aligned}
$$

3)
$$(\mathcal{L}_X(\omega) \wedge \omega')(X_1, X_2) + (\omega \wedge \mathcal{L}_X(\omega'))(X_1, X_2)$$
$$= \mathcal{L}_X(\omega)(X_1)\omega'(X_2) - \mathcal{L}_X(\omega)(X_2)\omega'(X_1)$$
$$+ \omega(X_1)\mathcal{L}_X(\omega')(X_2) - \omega(X_2)\mathcal{L}_X(\omega')(X_1)$$
$$= X(\omega(X_1))\omega'(X_2) - \omega([X, X_1])\omega'(X_2)$$
$$- X(\omega(X_2))\omega'(X_1) + \omega([X, X_2])\omega'(X_1)$$
$$+ \omega(X_1)X(\omega'(X_2)) - \omega(X_1)\omega'([X, X_2])$$
$$- \omega(X_2)X(\omega'(X_1)) + \omega(X_2)\omega'([X, X_1])$$
$$= X(\omega(X_1)\omega'(X_2) - \omega(X_2)\omega'(X_1))$$
$$- \omega([X, X_1])\omega'(X_2) + \omega(X_2)\omega'([X, X_1])$$
$$- \omega(X_1)\omega'([X, X_2]) + \omega([X, X_2])\omega'(X_1)$$
$$= X((\omega \wedge \omega')(X_1, X_2)) - (\omega \wedge \omega')([X, X_1], X_2)$$
$$- (\omega \wedge \omega')(X_1, [X, X_2])$$
$$= \mathcal{L}_X(\omega \wedge \omega')(X_1, X_2).$$

4) $\mathcal{L}_X(d\omega)(X_1, X_2)$
$$= X(d\omega(X_1, X_2)) - d\omega([X, X_1], X_2) - d\omega(X_1, [X, X_2])$$
$$= X\big(X_1(\omega(X_2))\big) - X\big(X_2(\omega(X_1))\big) - X(\omega([X_1, X_2]))$$
$$- [X, X_1](\omega(X_2)) + X_2(\omega([X, X_1])) + \omega([[X, X_1], X_2])$$
$$- X_1(\omega([X, X_2])) + [X, X_2](\omega(X_1)) + \omega([X_1, [X, X_2]]).$$

By the Lie-module axiom on one side and the Jacobi identity on the other side, we get

$$\mathcal{L}_X(d\omega)(X_1, X_2) = X_1(X(\omega(X_2)) - X_2(X(\omega(X_1)) - X(\omega([X_1, X_2]))$$
$$- X_1(\omega([X, X_2]) + X_2(\omega([X, X_1])) + \omega([X, [X_1, X_2]])$$
$$= X_1(\mathcal{L}_X\omega(X_2)) - X_2(\mathcal{L}_X\omega(X_1)) - \mathcal{L}_X\omega([X_1, X_2])$$
$$= d(\mathcal{L}_X\omega)(X_1, X_2).$$

5) $i_X(d\omega)(Y) + d(i_X\omega)(Y)$

$$= d\omega(X,Y) + d(\omega(X))(Y)$$
$$= X(\omega(Y)) - Y(\omega(X)) - \omega([X,Y]) + Y(\omega(X))$$
$$= \mathcal{L}_X\omega(Y).$$

6)
$$[L_X, L_Y](\omega)(Z) = L_X(L_Y(\omega))(Z) - L_Y(L_X(\omega))(Z)$$
$$= X\big(Y(\omega(Z))\big) - X(\omega([Y,Z])) - Y(\omega([X,Z]))$$
$$+ \omega([Y,[X,Z]]) - Y\big(X(\omega(Z))\big) + Y(\omega([X,Z]))$$
$$+ X(\omega([Y,Z])) - \omega([X,[Y,Z]])$$
$$= [X,Y](\omega(Z)) + \omega([Y,[X,Z]]) - \omega([X,[Y,Z]])$$
$$= [X,Y](\omega(Z)) - \omega([[X,Y],Z])$$
$$= \mathcal{L}_{[X,Y]}(\omega)(Z).$$

7)
$$[\mathcal{L}_X, i_Y](\omega) = \mathcal{L}_X(i_Y(\omega)) - i_Y(\mathcal{L}_X(\omega))$$
$$= \mathcal{L}_X(\omega(Y)) - \mathcal{L}_X(\omega)(Y)$$
$$= X(\omega(Y) - X(\omega(Y)) + \omega([X,Y])$$
$$= i_{[X,Y]}(\omega).$$

∎

6.1.3 Connections

In chapter 5, we defined not only pointwise connections (5.1.1 definition 1) but also global connections as some maps from $\mathcal{X}(M) \times \mathcal{X}(M)$ to $\mathcal{X}^2(M)$ (5.1.1 definition 2). But we constructed also a bijection between the set of global connections and the set of covariant derivations. These are global actions from $\mathcal{X}(M) \times \mathcal{X}(M)$ to $\mathcal{X}(M)$. And this concept can be generalized to general Lie objects.

Definition 8. *Let (A, L) be a Lie object and E an A-module. An action of (A, L) on E is a R-bilinear map*

$$\alpha : L \times E \to E : (X, v) \mapsto \alpha_X(v)$$

such that the derivation condition

$$\alpha_X(a \cdot v) = a \cdot \alpha_X(v) + X(a) \cdot v$$

is satisfied for all X in L, v in E and a in A.

Definition 9. A linear connection on E is an action of (A, L) on E which is A-linear in the variable X :

$$\alpha_{a.X}(v) = a \cdot \alpha_X(v).$$

As L is an A-module we have, in particular, a concept of connection on L.

We say that an action is flat (or is a Lie action) if $\forall X, Y \in L$ and $\forall v \in E$

$$\alpha_{[X,Y]}(v) = \alpha_X(\alpha_Y(v)) - \alpha_Y(\alpha_X(v)).$$

For example, the Lie bracket $[-] : L \times L \to L$ is a flat action of (A, L) on L. By the Koszul axiom this is an action and it is flat by the Jacobi identity. But this Lie bracket is not a connection.

Another triviality : the given action of L on A can be seen as a flat connection on A. (This is an action by the axiom of derivation ; it is flat by the axiom of Lie module ; and we have the A-linearity in X). If an A-module E is provide with a connection ∇, we can extend the exterior differential of scalar forms into a exterior covariant differential d_∇ of forms with values in E.

Definition 10. Let $\omega \in \Omega_A^p(L, E)$ and $X_0, \ldots, X_p \in L$. We define :

$$(d_\nabla \omega)(X_0, \ldots, X_p) = \sum_{i=0}^{p} (-1)^i \nabla_{X_i}(\omega(X_0, \ldots, \hat{X}_i, \ldots, X_p))$$

$$+ \sum_{0 \le i < j \le p} (-1)^{i+j} \omega([X_i, X_j], \ldots, \hat{X}_i, \ldots, \hat{X}_j, \ldots, X_p).$$

Proposition 7. If ω is in $\Omega_A^p(L, E)$, then $d_\nabla \omega$ is in $\Omega_A^{p+1}(L, E)$.

Proof: We only consider the case $p = 1$. We must show that

$$(d_\nabla \omega)(X_0, X_1) = \nabla_{X_0}(\omega(X_1)) - \nabla_{X_1}(\omega(X_0)) - \omega([X_0, X_1])$$

is A-multilinear and antisymmetrical. The antisymmetry is trivial. We show the A-linearity in X_0. We have

$$
\begin{aligned}
(d_\nabla \omega)(aX_0, X_1) &= a\nabla_{X_0}(\omega(X_1)) - \nabla_{X_1}(a\omega(X_0)) - \omega([aX_0, X_1]) \\
&= a\nabla_{X_0}(\omega(X_1)) - a\nabla_{X_1}(\omega(X_0)) - X_1(a)\omega(X_0) \\
&\quad - a\omega([X_0, X_1]) + X_1(a)\omega(X_0)
\end{aligned}
$$

by the definition of connections and the Koszul axiom. ∎

Proposition 8. d_∇ *is an antiderivation :*

$$
d_\nabla(\omega \wedge \omega') = d\omega \wedge \omega' + (-1)^p \omega \wedge \omega'.
$$

Proof: Similar to that of proposition 6 (1). ∎

Note that, if we take the trivial flat connection on A, we find again the exterior differential of scalar forms.

We however note that $d_\nabla \circ d_\nabla$ generally is not zero.

If ∇ is a connection on E, we can extend the Lie derivative to forms with values in E by

Definition 11. *We put*

$$
\begin{aligned}
L_X^\nabla(\omega)(X_1, \dots, X_p) &= \nabla_X(\omega(X_1, \dots, X_p)) \\
&\quad - \sum_{i=1}^p \omega(X_1, \dots, [X, X_i], \dots, X_p).
\end{aligned}
$$

Proposition 9. *We have*

$$
L_X^\nabla(\omega) = i_X(d_\nabla \omega) + d_\nabla(i_X \omega).
$$

Proof: By a computation as in proposition 6 (5). ∎

6.1.4   Extensions of connections and actions

Let (A, L) a Lie object, E and F two A-modules, α an action of (A, L) on E and β an action of (A, L) on F. We denote by $L_A^p(E, F)$ the A-module of A-multilinear maps from E^p to F. We shall construct an action $\beta(\alpha)$ of (A, L) on $L_A^p(E, F)$.

Definition 12. For $X \in L$, $U \in L_A^p(E, F)$ and $v_1, \ldots, v_p \in E$,

$$\beta(\alpha)_X(U)(v_1, \ldots, v_p) = \beta_X(U(v_1, \ldots, v_p))$$
$$- \sum_{i=1}^{p} U(v_1, \ldots, \alpha_X(v_i), \ldots, v_p).$$

Proposition 10.

1) $\beta(\alpha)$ is an action of (A, L) on $L_A^p(E, F)$.

2) $\beta(\alpha)_X(U)$ is antisymmetrical (resp. symmetrical) if U is antisymmetrical (resp. symmetrical).

3) If α and β are connection, $\beta(\alpha)$ is a connection.

4) If α and β are flat, $\beta(\alpha)$ is flat.

Proof:

1) We first observe that $\beta(\alpha)_X(U)$ is in $L_A^p(E, F)$. We get the A-linearity as follows

$$\beta(\alpha)_X(U)(a \cdot v_1, \ldots, v_p) = \beta_X(U(a \cdot v_1, \ldots, v_p))$$
$$- U(\alpha_X(a \cdot v_1), v_2, \ldots, v_p) - \sum_{i=2}^{p} U(av_1, \ldots, \alpha_X(v_i), \ldots, v_p)$$
$$= a \cdot \beta_X(U(v_1, \ldots, v_p)) + X(a)U(v_1, \ldots, v_p)$$
$$- a \cdot U(v_1, \ldots, v_p) - X(a)U(v_1, \ldots, v_p)$$
$$- a \sum_{i=2}^{p} U(v_1, \ldots, \alpha_X(v_i), \ldots, v_p)$$
$$= a \cdot \beta(\alpha)_X(U)(v_1, \ldots, v_p).$$

Then we show that $\beta(\alpha)$ is an action as, β being an action,

$$\beta(\alpha)_X(a \cdot U)(v_1, \ldots, v_p) = a \cdot \beta_X(U(v_1, \ldots, v_p))$$
$$+ X(a)U(v_1, \ldots, v_p) - \sum_{i=1}^{p} aU(v_1, \ldots, \alpha_X(v_i), \ldots, v_p)$$
$$= a \cdot \beta(\alpha)_X(U)(v_1, \ldots, v_p) + X(a) \cdot U(v_1, \ldots, v_p).$$

2) If U is symmetrical (or antisymmetrical) then clearly $\beta(\alpha)_X(U)$ is symmetrical (or antisymmetrical).

3) If α and β are connections, $\beta(\alpha)$ is also a connection. The A-linearity of $\beta(\alpha)_X$ in X comes from the A-linearity of β_X and α_X.

4) Let α and β be flat actions. We have

$$\beta(\alpha)_X(\beta(\alpha)_Y(U))(v_1,\ldots,v_p) = \beta_X(\beta_Y(U(v_1,\ldots,v_p))$$
$$- \sum_{i=1}^{p} \beta_X(U(v_1,\ldots,\alpha_Y(v_i),\ldots,v_p))$$
$$- \sum_{i=1}^{p} \beta_Y(U(v_1,\ldots,\alpha_X(v_i),\ldots,v_p))$$
$$+ \sum_{i\neq j} U(v_1,\ldots,\alpha_X(v_i),\ldots,\alpha_Y(v_j),\ldots,v_p)$$
$$+ \sum_{i} U(v_1,\ldots,\alpha_Y(\alpha_X(v_i)),\ldots,v_p).$$

Exchanging X and Y, we obtain by substraction

$$\Big(\beta(\alpha)_X(\beta(\alpha)_Y(U)) - \beta(\alpha)_Y(\beta(\alpha)_X(U))\Big)(v_1,\ldots,v_p)$$
$$= \beta_{[X,Y]}(U(v_1,\ldots,v_p)) - \sum_{i=1}^{p} U(v_1,\ldots,\alpha_{[X,Y]}(v_i),\ldots,v_p)$$

and $\beta(\alpha)$ is flat. ∎

In particular, if α is a connection on L and β a connection on E, $\beta(\alpha)$ is a connection on the A-modules of differential forms $\Omega_A^p(L, E)$.

As another particular case, if α is an action of (A, L) on E (resp. a connection on E or a flat action on A), $\alpha^* = \alpha(\alpha)$ is an action (resp. a connection or a flat action) on the A-module $\mathcal{L}_A(E)$ of the A-linear endomorphisms of E. We get

$$\alpha_X^*(U)(v) = \alpha_X(U(v)) - U(\alpha_X(v)).$$

6.2   Curvature of a connection on E and torsion of a connection on L

6.2.1   Curvature of a connection

Let (A, L) a Lie object, E an A-module and α an action of (A, L) on E. For $X, Y \in L$ we define

$$R_\alpha(X, Y) : E \to E$$

by

$$R_\alpha(X, Y)(v) = \alpha_X(\alpha_Y(v)) - \alpha_Y(\alpha_X(v)) - \alpha_{[X,Y]}(v).$$

Proposition 1. *For all action of* (A, L) *on* E, $R_\alpha(X, Y) : E \to E$ *is* A-*linear.*

Proof: We have

$$\alpha_X(\alpha_Y(a \cdot v)) = \alpha_X(a \cdot \alpha_Y(v)) + \alpha_X(Y(a) \cdot v)$$
$$= a \cdot \alpha_X(\alpha_Y(v)) + X(a) \cdot \alpha_Y(v) + Y(a) \cdot \alpha_X(v) + X(Y(a)) \cdot v$$

and, similarly,

$$\alpha_Y(\alpha_X(a \cdot v)) = a \cdot \alpha_Y(\alpha_X(v)) + Y(a)\alpha_X(v)$$
$$+ X(a) \cdot \alpha_Y(v) + Y(X(a)) \cdot v$$
$$\alpha_{[X,Y]}(a \cdot v) = a \cdot \alpha_{[X,Y]}(v) + [X, Y](a) \cdot v.$$

Thus, by the Lie-module axiom,

$$R_\alpha(X, Y)(a \cdot v) = a \cdot R_\alpha(X, Y)(v).$$

■

Proposition 2. *If* ∇ *is a connection on* E, $R_\nabla : L \times L \to \mathcal{L}_A(E) :$ $(X, Y) \mapsto R_\nabla(X, Y)$ *is a 2-form with values in* $\mathcal{L}_A(E)$.

Proof: We immediately get antisymmetry. We show the A-linearity in X :

$$R_\nabla(aX, Y)(v) = \nabla_{aX}(\nabla_Y(v)) - \nabla_Y(\nabla_{aX}(v)) - \nabla_{[aX,Y]}(v).$$

The Koszul's axiom give

$$[aX, Y] = a[X, Y] - Y(a)X$$

and, as ∇_X is linear in X, we obtain

$$\begin{aligned} R_\nabla(aX, Y)(v) &= a\nabla_X(\nabla_Y(v)) - \nabla_Y(a \cdot \nabla_X(v)) \\ &- a\nabla_{[X,Y]}(v) + Y(a)\nabla_X(v). \end{aligned}$$

By the derivation condition of actions (6.1 definition 8), we have

$$\nabla_Y(a \cdot \nabla_X(v)) = a \cdot \nabla_Y(\nabla_X(v)) + Y(a)\nabla_X(v).$$

We have thus

$$R_\nabla(aX, Y)(v) = a \cdot R_\nabla(X, Y)(v).$$

∎

In the sequel, we suppose the connection ∇ on E to be settled and we write R for R_∇.

Observe that the 2-form R with values in $\mathcal{L}_A(E)$ can be seen as an A-linear map from E to $\Omega^2(L, E)$. We simply take $R(v)(X, Y) = R(X, Y)(v)$. We can extend that into an operator of degree two on the forms with values in E.

Definition 1. *We define*

$$R : \Omega^p(L, E) \to \Omega^{p+2}(L, E)$$

by :

$$R(\omega)(X_0, X_1, \ldots, X_{p+1}) =$$
$$\sum_{0 \leq i < j \leq p+1} (-1)^{i+j+1} R(X_i, X_j)(\omega(X_0, \ldots, \hat{X}_i, \ldots, \hat{X}_j, \ldots, X_{p+1})).$$

We must observe that $R(\omega)$ is A-multilinear and antisymmetrical. But it is obvious.

We noted (6.1.3) that the exterior covariant differential d_∇ generally is not of square zero. In fact we have

Proposition 3. *As maps from $\Omega^p(L, E)$ into $\Omega^{p+2}(E)$ we have*

$$d_\nabla \circ d_\nabla = R.$$

Proof: We will give an explicit computation in spite of its straightforward character.

To divide the difficulty, we begin with the case $p = 1$. First we have :

$$d_\nabla(d_\nabla(\omega))(X_0, X_1, X_2) =$$
$$\nabla_{X_0}(d_\nabla(\omega)(X_1, X_2) - \nabla_{X_1}(d_\nabla(\omega)(X_0, X_2)) + \nabla_{X_2}(d_\nabla(\omega)(X_0, X_1))$$
$$- d_\nabla(\omega)([X_0, X_1], X_2) + d_\nabla(\omega)([X_0, X_2], X_1) - d_\nabla(\omega)([X_1, X_2], X_0)$$
$$= \nabla_{X_0}\Big(\nabla_{X_1}(\omega(X_2))\Big) - \nabla_{X_0}\Big(\nabla_{X_2}(\omega(X_1))\Big) - \nabla_{X_0}(\omega([X_1, X_2]))$$
$$- \nabla_{X_1}\Big(\nabla_{X_0}(\omega(X_2))\Big) + \nabla_{X_1}\Big(\nabla_{X_2}(\omega(X_0))\Big) + \nabla_{X_1}(\omega([X_0, X_2]))$$
$$+ \nabla_{X_2}\Big(\nabla_{X_0}(\omega(X_1))\Big) - \nabla_{X_2}\Big(\nabla_{X_1}(\omega(X_0))\Big) - \nabla_{X_2}(\omega([X_0, X_1]))$$
$$- \nabla_{[X_0, X_1]}(\omega(X_2)) + \nabla_{X_2}(\omega([X_0, X_1])) + \omega([[X_0, X_1], X_2])$$
$$+ \nabla_{[X_0, X_2]}(\omega(X_0)) - \nabla_{X_1}(\omega([X_0, X_2])) - \omega([[X_0, X_2], X_1])$$
$$- \nabla_{[X_1, X_2]}(\omega(X_0)) + \nabla_{X_0}(\omega([X_1, X_2])) + \omega([[X_1, X_2], X_0]).$$

The Jacobi identity permit the annulation of the terms with two brackets. The terms with simple bracket inside of ω disappear. We have thus

$$d_\nabla(d_\nabla(\omega))(X_0, X_1, X_2) =$$
$$\nabla_{X_0}\nabla_{X_1}(\omega(X_2)) - \nabla_{X_1}\nabla_{X_0}(\omega(X_2)) - \nabla_{[X_0, X_1]}(\omega(X_2))$$
$$- \nabla_{X_0}\nabla_{X_2}(\omega(X_1)) + \nabla_{X_2}\nabla_{X_0}(\omega(X_1)) + \nabla_{[X_0, X_2]}(\omega(X_1))$$
$$+ \nabla_{X_1}\nabla_{X_2}(\omega(X_0)) - \nabla_{X_2}\nabla_{X_1}(\omega(X_0)) - \nabla_{[X_1, X_2]}(\omega(X_0))$$
$$= R(X_0, X_1)(\omega(X_2)) - R(X_0, X_2)(\omega(X_1)) + R(X_1, X_2)(\omega(X_0))$$
$$= R(\omega)(X_0, X_1, X_2).$$

For general p, the difficulties are only of writing. But, to see as this work, we also give the explicit computation for $\omega \in \Omega^p(L, E)$.

$$d_\nabla(d_\nabla(\omega))(X_0, \ldots, X_{p+1})$$

$$= \sum_{i=0}^{p+1} (-1)^i \nabla_{X_i}((d_\nabla \omega)(\dots, \hat{X}_i, \dots))$$

$$+ \sum_{i<j} (-1)^{i+j} (d_\nabla \omega)([X_i, X_j], \dots, \hat{X}_i, \dots, \hat{X}_j, \dots)$$

$$= \sum_{i,j} \epsilon_{ij} \nabla_{X_i} \left(\nabla_{X_j} (\omega(\dots, \hat{X}_i, \dots, \hat{X}_j, \dots)) \right) \tag{1}$$

(where $\epsilon_{ij} = (-1)^{i+j}$ if $j < i$, $\epsilon_{ij} = (-1)^{i+j+1}$ if $j > i$, $\epsilon_{ii} = 0$)

$$+ \sum_i \sum_{r<s} \epsilon_{irs} \nabla_{X_i} (\omega([X_r, X_s], \dots, \hat{X}_r, \dots, \hat{X}_s, \dots, \hat{X}_i, \dots)) \tag{2}$$

(where $\epsilon_{irs} = (-1)^{i+r+s}$ if $r < s < i$ or $i < r < s$, $\epsilon_{irs} = (-1)^{i+r+s+1}$ if $r < i < s$ and $\epsilon_{irs} = 0$ if $r = i$ or $s = i$)

$$+ \sum_{i<j} (-1)^{i+j} \nabla_{[X_i, X_j]} (\omega(\dots, \hat{X}_i, \dots, \hat{X}_j, \dots) \tag{3}$$

$$+ \sum_{i<j} \sum_k \delta_{ijk} \nabla_{X_k} (\omega([X_i, X_j], \dots, \hat{X}_i, \dots, \hat{X}_j, \dots, \hat{X}_k, \dots)) \tag{4}$$

(where $\delta_{ijk} = (-1)^{i+j+k+1}$ if $k < i$ or $j < k$, $\delta_{ijk} = (-1)^{i+j+k}$ if $i < k < j$ and $\delta_{ijk} = 0$ if $k = i$ or $k = j$)

$$+ \sum_{i<j} \sum_r \gamma_{ijr} \omega([[X_i, X_j], X_r], \dots, \hat{X}_i, \dots, \hat{X}_j, \dots, \hat{X}_r, \dots) \tag{5}$$

(where $\gamma_{ijr} = (-1)^{i+j+r}$ if $r < i$ or $j < r$, $\gamma_{ijr} = (-1)^{i+j+r+1}$ if $i < r < j$ and $\gamma_{ijr} = 0$ if $r = i$ or $r = j$)

$$+ \sum_{i<j} \sum_{r<s} \beta_{ijrs} \omega([X_r, X_s], [X_i, X_j], \dots, \hat{X}_i, \dots, \hat{X}_j, \dots, \hat{X}_r, \dots, \hat{X}_s, \dots)$$
$$\tag{6}$$

(where $\beta_{ijrs} = (-1)^{i+j+r+s}$ if $r < s < i < j$ or $r < i < j < s$ or $i < j < r < s$ or $i < r < s < j$, $\beta_{ijrs} = (-1)^{i+j+r+s+1}$ if $r < i < s < j$ or $i < r < j < s$ and $\beta_{ijrs} = 0$ if $r = i$ or $r = j$ or $s = i$ or $s = j$).

As $\delta_{ijr} = -\epsilon_{kij}$, $(2) + (4) = 0$. By the Jacobi identity, the sum (5) is zero. By antisymmetry the sum (6) is zero. It remains $(1) + (3)$, that is :

$$d_\nabla(d_\nabla\omega)(X_0,\ldots,X_{p+1}) =$$
$$\sum_{i<j}(-1)^{i+j+1}\nabla_{X_i}\left(\nabla_{X_j}(\omega(\ldots,\hat{X}_i,\ldots,\hat{X}_j,\ldots))\right)$$
$$+\sum_{i<j}(-1)^{i+j}\nabla_{X_j}\left(\nabla_{X_i}(\omega(\ldots,\hat{X}_i,\ldots,\hat{X}_j,\ldots))\right)$$
$$+\sum_{i<j}(-1)^{i+j}\nabla_{[X_i,X_j]}(\omega(\ldots,\hat{X}_i,\ldots,\hat{X}_j,\ldots))$$
$$=\sum_{i<j}(-1)^{i+j+1}R(X_i,X_j)(\omega(\ldots,\hat{X}_i,\ldots,\hat{X}_j,\ldots))$$
$$= R(\omega)(X_0,\ldots,X_{p+1}).$$

∎

6.2.2 Bianchi identity

It is possible to derive *Bianchi identity* from proposition 3. But a direct computation is also available and we prefer this.

The connection ∇ extends into a connection ∇^* on $\mathcal{L}_A(E)$, as indicated at the end of section 6.1.3 :

$$\nabla^*_X(U)(v) = \nabla_X(U(v)) - U(\nabla_X(v)).$$

The curvature R is viewed as a 2-form with values in $\mathcal{L}_A(E)$. We can thus formulate the following proposition.

Proposition 4. (Bianchi identity). *For any connection ∇ on E, we have*

$$d_{\nabla^*}(R) = 0.$$

Proof: The definition of exterior covariant differential (6.1.3, definition 10) gives to us :

$$d_{\nabla^*}(R)(X_0,X_1,X_2)(v) = \nabla^*_{X_0}(R(X_1,X_2))(v)$$
$$- \nabla^*_{X_1}(R(X_0,X_2))(v) + \nabla^*_{X_2}(R(X_0,X_1))(v)$$
$$- R([X_0,X_1],X_2)(v) + R([X_0,X_2],X_1)(v) - R([X_1,X_2],X_0)(v)$$

which, by the definition of ∇^*, gives

$$\nabla_{X_0}(R(X_1,X_2)(v)) - R(X_1,X_2)(\nabla_{X_0}(v))$$
$$- \nabla_{X_1}(R(X_0,X_2)(v)) + R(X_0,X_2)(\nabla_{X_1}(v))$$
$$+ \nabla_{X_2}(R(X_0,X_1)(v)) - R(X_0,X_1)(\nabla_{X_2}(v))$$
$$- R([X_0,X_1],X_2)(v) + R([X_0,X_2],X_1)(v) - R([X_1,X_2],X_0)(v).$$

We substitute the expression of R and all the terms cancel two by two, with the exception of

$$\nabla_{[[X_0,X_1],X_2]}(v) - \nabla_{[[X_0,X_2],X_1]}(v) + \nabla_{[[X_1,X_2],X_0]}(v).$$

But this equals zero by Jacobi identity. ∎

6.2.3 Torsion of a connection on L

Here we only consider a connection on L.

Definition 2. *Let $\nabla : L \times L \to L$ be a connection on L. The torsion of ∇ is the map $T : L \times L \to L$ given by*

$$T(X,Y) = \nabla_X Y - \nabla_Y X - [X,Y].$$

Proposition 5. *If ∇ is a connection on L, its torsion is a 2-form with values in L.*

Proof: That T is antisymmetrical is trivial. We prove the A-linearity in X.

$$\begin{aligned}
T(a \cdot X, Y) &= \nabla_{a \cdot X} Y - \nabla_Y(aX) - [aX,Y]\\
&= a \cdot \nabla_X Y - a\nabla_Y(X) - Y(a)X - a[X,Y] + Y(a)X\\
&= a T(X,Y).
\end{aligned}$$

∎

The following result is often called the first Bianchi identity.

Proposition 6. *Let ∇ be a connection on L, R its curvature and T its torsion. For all X,Y and Z in L,*

$$\sum_0 R(X,Y)(Z) = \sum_0 T(T(X,Y),Z) + \sum_0 (\nabla_X T)(Y,Z)$$

(where $\sum_0$ denotes the sum on the circular permutations of (X,Y,Z)).

Proof: By definition 12 in 6.1.4, we have

$$(\nabla_X T)(Y,Z) = \nabla_X(T(Y,Z)) - T(\nabla_X Y, Z) - T(Y, \nabla_X Z).$$

Using the definition of T, the righthandside of the thesis can be written :

$$\sum_0 T(\nabla_X Y, Z) - \sum_0 T(\nabla_Y X, Z) - \sum_0 T([X,Y],Z)$$
$$+ \sum_0 \nabla_X(T(Y,Z)) - \sum_0 T(\nabla_X Y, Z) - \sum_0 T(Y, \nabla_X Z).$$

As T is antisymmetrical, this reduces to

$$\sum_0 \nabla_X(T(Y,Z)) - \sum_0 T([X,Y],Z).$$

Again using the definition of T, we get

$$\sum_0 \nabla_X \nabla_Y Z - \sum_0 \nabla_X \nabla_Z Y - \sum_0 \nabla_X([Y,Z])$$
$$- \sum_0 \nabla_{[X,Y]} Z + \sum_0 \nabla_Z([X,Y]) - \sum_0 [[X,Y],Z].$$

The last term is zero by Jacobi's identity. The circular sums $\sum_0 \nabla_Z([X,Y]) - \sum_0 \nabla_X([Y,Z])$ cancel. What remains is only, as asked, $\sum_0 R(X,Y)(Z)$. ∎

To a connection ∇ on L, we associate ∇^* on $\mathcal{L}_A(L)$ and the Bianchi identity (proposition 4) is written

$$\sum_0 \nabla^*_X(R(Y,Z)) = \sum_0 R([X,Y],Z).$$

But, it is also possible to extend ∇ to $\Omega^2(L, \mathcal{L}_A(L))$. We denote $\tilde{\nabla}$ this extension. We obtain another version of the Bianchi identity.

Proposition 7.

$$\sum_0 (\tilde{\nabla}_X R)(Y, Z) = \sum_0 R(X, T(Y, Z)).$$

Proof: The lefthandside is

$$\sum_0 (\tilde{\nabla}_X R)(Y, Z) = \sum_0 \nabla_X \circ R(Y, Z) - \sum_0 R(Y, Z) \circ \nabla_X$$
$$- \sum_0 R(\nabla_X Y, Z) - \sum_0 R(Y, \nabla_X Z).$$

By the definition of ∇^* :

$$\nabla_X^*(R(Y, Z)) = \nabla_X \circ R(Y, Z) - R(Y, Z) \circ \nabla_X$$

and the lefthandside becomes

$$\sum_0 (\tilde{\nabla}_X R)(Y, Z) = \sum_0 \nabla_X^*(R(Y, Z)) - \sum_0 R(\nabla_X Y, Z)$$
$$- \sum_0 R(Y, \nabla_X Z)$$

or, by the Bianchi identity

$$\sum_0 (\tilde{\nabla}_X R)(Y, Z) = \sum_0 R([X, Y], Z) - \sum_0 R(\nabla_X Y, Z) - \sum_0 R(Y, \nabla_X Z).$$

But the righthandside is

$$\sum_0 R(X, T(Y, Z)) = \sum_0 R(X, \nabla_Y, Z)$$
$$- \sum_0 R(X, \nabla_Z(Y)) - \sum_0 R(X, [Y, Z])$$

and the computational proof is complete. ∎

If the connection ∇ is torsion free, we trivially again find the usual formulas :

Proposition 8. *If the torsion T is null, we have :*

$$\sum_0 R(X, Y)(Z) = 0$$
$$\sum_0 (\tilde{\nabla}_X R)(Y, Z) = 0.$$

We close this section in pointing out the following results.

Proposition 9. *If ∇ is a torsion free connection, and ψ is a 2-form,*

$$2d\psi(X_1, X_2, X_3) = \sum_{\sigma \in S_3} \epsilon_\sigma (\nabla_{X_{\sigma_1}} \psi)(X_{\sigma_2}, X_{\sigma_3})$$

where

$$(\nabla_X \psi)(Y, Z) = X(\psi(Y, Z)) - \psi(\nabla_X Y, Z) - \psi(Y, \nabla_X Z).$$

Proof: Transfer the definition of $\nabla_{X_i}\psi$ in the righthandside, use that ψ is antisymmetrical and $\nabla_X Y - \nabla_Y X = [X, Y]$ (as the torsion is zero) and you obtain $2d\psi(X_1, X_2, X_3)$. ∎

6.3 Weil's characteristic homomorphism

6.3.1 Algebraic preliminaries

First we fix some notations. We always have a Lie object (A, L). Let F be an A-module and $\Lambda \in \mathcal{L}^p(F, A)$. Let Λ operates on differential forms with values in F, transforming these in a scalar form. More precisely :

Definition 1. *Let $\omega_1 \in \Omega^{q_1}(L, F), \ldots, \omega_p \in \Omega^{q_p}(L, F)$ be forms with values in F. We define*

$$(\Lambda | \omega_1, \ldots, \omega_p)(X_1, \ldots, X_q) =$$

$$\frac{1}{q_1! \ldots q_p!} \sum_{\sigma \in S_q} \epsilon_\sigma \Lambda(\omega_1(X_{\sigma(1)}, \ldots, X_{\sigma(q_1)}),$$

$$\ldots, \omega_p(X_{\sigma(q_1 + \ldots + q_{p-1}+1)}, \ldots, X_{\sigma(q)}))$$

(where $q = q_1 + \ldots + q_p$).

Up to an integer factor $(\Lambda | \omega_1, \ldots, \omega_p)$ is the antisymmetrized form coming from $\Lambda(\omega_1, \ldots, \omega_p)$. We thus have

Proposition 1. $(\Lambda | \omega_1, \ldots, \omega_p)$ *is a scalar q-form.* ∎

We can extend that construction, replacing $\Lambda \in \mathcal{L}^p(F, A)$ by 2-forms with values in $\mathcal{L}^p(F, A)$, so obtaining a scalar $(q + r)$-form. We explicit only the case $r = 1$.

Definition 2. *Let Γ be in $\Omega^1(L, \mathcal{L}^p(F, A))$, we define $(\Gamma|\omega_1, \ldots, \omega_p)$ as the $(q+1)$-form given by*

$$(\Gamma|\omega_1, \ldots, \omega_p)(X_0, X_1, \ldots, X_q) =$$

$$\frac{1}{q_1! \ldots q_p!} \sum_{\tau \in S_{q+1}} \epsilon_\tau \Gamma(X_{\tau(0)})(\omega_1(X_{\tau(1)}, \ldots, X_{\tau(q_1)}), \ldots, \omega_p(\ldots, X_{\tau(q)})).$$

Let ∇^* a connection on F. We denote by I the trivial flat connection on A (cf. 6.1.3) and we consider the connection $I(\nabla^*)$ on $\mathcal{L}(F, A)$. By the definition 12 of 6.1.4 we immediately have

$$I(\nabla^*)_X(\Lambda)(U_1, \ldots, U_p) = X(\Lambda(U_1, \ldots, U_p))$$
$$- \sum_{i=1}^{p} \Lambda(U_1, \ldots, \nabla_X^*(U_i), \ldots, U_p).$$

We simply denote by $\tilde{\nabla}(\Lambda)$ the 1-form with values in $\mathcal{L}^p(F, A)$ given by :

$$\tilde{\nabla}(\Lambda)(X) = I(\nabla^*)_X(\Lambda).$$

6.3.2 A derivation formula

A tedious but crucial computation is necessary for the following natural proposition which is a kind of derivation formula.

Proposition 2. *Let F be an A-module with a connection ∇^*, $\Lambda \in \mathcal{L}^p(F, A)$, $\omega_1 \in \Omega^{q_1}(L, F), \ldots, \omega_p \in \Omega^{q_p}(L, F)$. Let $s_i = q_1 + \ldots + q_i$. We have :*

$$d((\Lambda|\omega_1, \ldots, \omega_p)) = (\tilde{\nabla}(\Lambda)|\omega_1, \ldots, \omega_p)$$
$$+ \sum_{i=1}^{p}(-1)^{s_{i-1}}(\Lambda|\omega_1, \ldots, d_{\nabla^*}(\omega_i), \ldots, \omega_p).$$

Proof:

a) By the description of d (6.1.2 proposition 5) and the definition 1 we have :

$$d((\Lambda|\omega_1, \ldots, \omega_p))(X_0, \ldots, X_{s_p})$$

$$= \sum_{i=0}^{s_p} \sum_{\sigma} (-1)^i \epsilon_\sigma \frac{1}{q_1! \dots q_p!} X_i \left(\Lambda(\omega_1(X_{\sigma(1)}), \dots, X_{\sigma(s_1)}), \right.$$

$$\left. \dots, \omega_p(X_{\sigma(s_{p-1}+1)}, \dots, X_{\sigma(s_p)})) \right)$$

$$+ \sum_{0 \le i < j \le s_p} \sum_{\tau} (-1)^{i+j} \epsilon_\tau \frac{1}{q_1! \dots q_p!} \Lambda(\omega_1(Y_{\tau(1)}, \dots, Y_{\tau(s_1)}),$$

$$\dots, \omega_p(Y_{\tau(s_{p-1}+1)}, \dots, Y_{\tau(s_p)})) \right)$$

where σ is in the set of bijections from $\{1, \dots, s_p\}$ onto $\{0, \dots, \hat{i}, \dots, s_p\}$; τ is in the set of bijections from $\{1, \dots, s_p\}$ onto $\{-1, 0, \dots, \hat{i}, \dots, \hat{j}, \dots, s_p\}$; $Y_{-1} = [X_i, X_j]$ and $Y_k = X_k$ for $k \ne -1$. We write the two sums of the expression as

$$\Sigma A_{i\sigma} + \Sigma B_{ij\tau}.$$

b) We compute the first term of the righthandside. As $\tilde{\nabla}(\Lambda)(X) = I(\nabla^*)_X(\Lambda)$ we have, by definition 2,

$$(\tilde{\nabla}(\Lambda)|\omega_1, \dots, \omega_p)(X_0, \dots, X_{s_p})$$

$$= \sum_{\tau \in S_{s_p+1}} \frac{1}{q_1! \dots q_p!} \epsilon_\tau X_{\tau(0)} (\Lambda(\omega_1(X_{\tau(1)}, \dots, X_{\tau(s_1)}),$$

$$\dots, \omega_p \left(X_{\tau(s_{p-1}+1)}, \dots, X_{\tau(s_p)} \right)))$$

$$- \sum_{\tau \in S_{s_p+1}} \sum_{i=1}^{p} \frac{1}{q_1! \dots q_p!} \epsilon_X \Lambda(\omega_1(X_{\tau(1)}, \dots, X_{\tau(s_1)}),$$

$$\dots, \nabla^*_{X_{\tau(0)}}(\omega)(X_{\tau(s_i-1+1)}, \dots, X_{\tau(s_i)}),$$

$$\dots, \omega_p(X_{\tau(s_{p-1}+1)}, \dots, X_{\tau(s_p)})).$$

We note the two sums by : $\Sigma C_\tau - \Sigma D_{\tau i}$.

c) Immediately we observe that $\Sigma A_{i\sigma} = \Sigma C_\tau$. To see that, we associate to a bijection τ of $(0, \dots, s_p)$ onto itself, the pair (i, σ) when $i = \tau(0)$ and σ is induced by τ. We have $A_{i\sigma} = C_\tau$ because $\epsilon_\sigma = (-1)^i \epsilon_\tau$.

d) For the calculation of the second term of the righthandside we use definition 1 and the definition of exterior covariant differential

(6.1.3, definition 10). We obtain :

$$\sum_{i=1}^{p} \sum_{\tau \in S_{s_p+1}} \sum_{r=0}^{q_i} \frac{1}{q_1! \dots q_p!} \frac{1}{q_i+1} (-1)^{s_i-1} \epsilon_\tau (-1)^r$$

$$\wedge (\omega_1(X_{\tau(0)}, \dots, X_{\tau(s_1-1)}), \dots,$$

$$\nabla^*_{X_{\tau(s_{i-1}+r)}} (\omega_i(X_{\tau(s_{i-1})}, \dots, \hat{X}_{\tau(s_{i-1}+r)}, \dots, X_{\tau(s_i)}),$$

$$\dots, \omega_p(X_{\tau(s_{p-1}+1)}, \dots, X_{\tau(s_p)})))$$

$$+ \sum_{i=1}^{p} \sum_{\tau \in S_{s_p+1}} \sum_{0 \le r < s \le s_p} \frac{1}{q_1! \dots q_p!} \frac{1}{q_i+1} (-1)^{s_i-1} \epsilon_\tau$$

$$(-1)^{r+s} \wedge (\omega_1(X_{\tau(0)}, \dots, X_{\tau(q_1-1)}), \dots,$$

$$\omega_i([X_{\tau(s_{i-1}+r)}, X_{\tau(s_{i-1}+s)}], \dots, \hat{X}_{\tau(s_{i-1}+r)}, \dots,$$

$$\hat{X}_{\tau(s_{i-1}+s)}, \dots, X_{\tau(s_i)}), \dots, \omega_p(X_{\tau(s_{p-1}+1)}, \dots, X_{\tau(s_p)}).$$

We denote these sums by $\Sigma E_{\tau i r} + \Sigma F_{\tau i r s}$.

e) $\Sigma E_{\tau i r} = \Sigma D_{\tau i}$. For, let $\tau \in S_{s_p+1}$, $1 \le i \le p$ and $0 \le r \le q_i$. We associate to (τ, i, r) the pair (τ', i) where $\tau'(0) = \tau(s_{i-1}+r)$, $\tau'(k) = \tau(k-1)$ for $1 \le k \le s_{i-1}+r$ and $\tau'(k) = \tau(k)$ for $s_{i-1}+r+1 \le k \le s_p$. We have $\epsilon_{\tau'} = (-1)^{s_i-1+r} \epsilon_\tau$ and we so obtain q_i+1 terms equal to $D_{\tau' i}$ and thus the announced equality.

f) To achieve the proof it remain to see that $\Sigma B_{ijr} = \Sigma F_{\tau i r s}$. We fix k with $1 \le k \le p$, we fix i and j with $1 \le i < j \le s_p$. We fix a sequence $\underline{u} = (u_1, \dots, u_{s_{k-1}}, u_{s_k+1}, \dots, u_{s_p})$ of $s_p - q_k$ indices. Among the q_k indices which are not in $\underline{u}$ we fix $q_k - 1$ indices $\underline{v} = (v_1, v_2, \dots, v_{q_k-1})$. In ΣB_{ijr} we only retain the terms, designated by $B_\tau^{k,i,j,\underline{u},\underline{v}}$, determinated by the bijections τ from $\{1, \dots, s_p\}$ onto $\{-1, 0, \dots, \hat{i}, \dots, \hat{j}, \dots, s_p\}$ such that $\tau(1) = u, \dots, \tau(s_{k-1}) = u_{s_{k-1}}, \tau(s_k+1) = u_{s_k+1}, \dots, \tau(s_p) = u_p$ and such that there exists an indice t $(1 \le t \le q_k)$ with

$$\tau(s_{k-1}+1) = v_1,$$

$$\dots,$$

$$\tau(s_{k-1}+t-1) = v_{t-1},$$
$$\tau(s_{k-1}+t) = -1,$$

$$\tau(s_{k-1} + t + 1) = v_t,$$

$$\cdots,$$

$$\tau(s_k) = v_{q_k} - 1.$$

So it remains only q_k possibilities for τ, one for each t ($1 \le t \le q_k$). We denote these by $\tau_1, \tau_2, \ldots, \tau_{q_k}$. But, for every t,

$$B_{\tau_t}^{k,i,j,\underline{u},\underline{v}} = B_{\tau_1}^{k,i,j,\underline{u},\underline{v}}$$

because the way to get τ_1 from τ_t is to bring the term with indice $-1 = \tau(s_{k-1} + t)$ in the position of indice $s_{k-1} + 1$. So we induce a factor $(-1)^{t-1}$ since ω_k is antisymmetrical. But moreover the change of signature from τ_t to τ, also is $(-1)^{t-1}$. So the sum of the $B_{ij\tau}$ corresponding to the choice of $(k, i, j, \underline{u}, \underline{v})$ reduces to

$$q_k \cdot B_{\tau_1}^{k,i,j,\underline{u},\underline{v}}$$

or, explicitly, to

$$q_k \cdot (-1)^{i+j} \epsilon_{\tau_1} \frac{1}{q_1! \ldots q_p!} \Lambda(\omega(X_{u_1}, \ldots, X_{u_{q_1}}),$$
$$\ldots, \omega_k([X_i, X_j], X_{v_1}, \ldots, X_{v_{q_k}} - 1),$$
$$\ldots, \omega_p(X_{u_{s_p-1}+1}, \ldots, X_{u_{s_p}})).$$

As well, in $\Sigma F_{\tau irs}$ we only retain the terms $F_{\tau rs}^{k,i,j,\underline{u},\underline{v}}$ determinated by the (r, s) with $0 \le r < s \le q_k$ and τ such that

$$\tau(0) = u_1,$$

$$\cdots,$$

$$\tau(s_{k-1} - 1) = u_{s_k-1},$$
$$\tau(s_{k-1}) = v_1,$$

$$\cdots,$$

$$\tau(s_{k-1} + r - 1) = v_r,$$
$$\tau(s_{k-1} + r) = i \text{ or } j,$$
$$\tau(s_{r-1} + r + 1) = v_{r+1},$$

$$\cdots,$$

$$\tau(s_{k-1} + s - 1) = v_{s-1},$$
$$\tau(s_{k-1} + s) = j \text{ or } i,$$
$$\tau(s_{k-1} + s + 1) = v_s,$$
$$\cdots,$$
$$\tau(s_k) = v_{q_k} - 1,$$
$$\tau(s_k + 1) = u_{s_k+1},$$
$$\cdots,$$
$$\tau(s_p) = u_{s_p}.$$

Finally such τ are determinated by pairs (r, s) with $0 \le r < s \le q_k$ and by the choice $\tau(s_{k+1} + r) = i$ (and thus $\tau(s_{k+1} + s) = j$) or $\tau(s_{k-1} + r) = j$ (and thus $\tau(s_{k-1} + s) = i$). In the first case we note this τ as $\tau_{r,s}$ and in the second case as $\tau'_{r,s}$. Their amount is $2 \cdot (q_k \cdot (q_k + 1))/2$. We then observe that all these terms correspond to $B^{k,i,j,u,v}_{\tau_1}$ and the proposition is proved.

∎

6.3.3 The Weil homomorphism

The preceding result shall be apply when $F = \mathcal{L}_A(E)$. Let ∇ be a connection on E and ∇^* the corresponding connection in $\mathcal{L}_A(E)$ (cf. 6.1.4). For any $\Lambda \in \mathcal{L}^p(\mathcal{L}_A(E), A)$, $\tilde{\nabla}(\Lambda)$ is a 1-form with values in $\mathcal{L}^p(\mathcal{L}_A(E), A)$. We also recall that the curvature R of the connection ∇, can be viewed as a 2-form with values in $\mathcal{L}_A(E)$.

Definition 3. *The Weil map*

$$W_1 : \mathcal{L}^p(\mathcal{L}_A(E), A) \to \Omega^{2p}(L)$$

is the map given by

$$W_1(\Lambda) = (\Lambda | R, R, \dots, R).$$

Proposition 3. *For every Λ in $\mathcal{L}^p(\mathcal{L}_A(E), A)$*

$$d(W_1(\Lambda)) = (\tilde{\nabla}(\Lambda) | R, \dots, R).$$

Proof: This comes from proposition 2 and the Bianchi identity $d_{\nabla^*}(R) = 0$. ∎

Definition 4. *If ∇ is a connection on E, we say $v \in E$ is ∇-parallel if $\nabla(v) = 0$.*

So $\Lambda \in \mathcal{L}^p(\mathcal{L}_A(E), A)$ is parallel if $\tilde{\nabla}(\Lambda) = 0$. Proposition 3 immediately gives

Proposition 4. Λ *is parallel if $d(W_1(\Lambda)) = 0$.* ∎

We denote by $\mathcal{L}^p_\nabla$, the algebra generated by parallel Λ. The form $W_1(\Lambda)$ is closed for Λ in $\mathcal{L}^p_\nabla$. Thus we get a map, in fact a homomorphism, from $\mathcal{L}^p_\nabla$ in the De Rham cohomology $H^{2p}(E; A)$.

This map $W : \mathcal{L}^p_\nabla \to H^{2p}(E, A)$ is the *Weil characteristic homomorphism*.

6.4 Commented bibliography

Global aspects of connections and differential forms (including the Weil caracteristic homomorphism) are introduced by Lavendhomme in [42] (1991) ; we note here that this is similar to the classical view. See for example the work of Koszul [39] (1950) (there are many classical works in the same spirit, for example, the Cartan formulas are in [11](1951), and perhaps in any book on differential geometry since). But the synthetic point of view remains more general and more simple, at least for generalities.

Chapter 7

On the algebra of the geometry of mechanics

7.1 Structured Lie objects

7.1.1 The Riemannian case

First we define the fundamental concept of Riemannian structure. We must clarify the notion of non-degenerate map, because, in our logic, the function of negation is tricky. Let A be an R-algebra and E an A-module. A A-bilinear map

$$h : E \times E \to A$$

is non-degenerate if the two partial maps

$$h_1 : E \to \mathcal{L}_A(E, A) : v \mapsto h(v, -)$$

$$h_2 : E \to \mathcal{L}_A(E, A) : v \mapsto h(-, v)$$

are inversible.

We have three concepts of *Riemannian structure*.

Definition 1.

1) *Let M be a microlinear object. A pointwise Riemannian structure on M is the datum for every $x \in M$ of a map*

$$g_x : T_x M \times T_x M \to R$$

R-bilinear, symmetric and nondegenerate.

2) A global Riemannian structure on M is a map

$$g : \mathcal{X}(M) \times \mathcal{X}(M) \to R^M$$

R^M-*bilinear, symmetric and non-degenerate.*

3) Let (A, L) be a Lie object. A Riemannian structure on (A, L) is a map

$$g : L \times L \to A$$

A-bilinear, symmetric and non-degenerate.

Obviously, a pointwise Riemannian structure induces a global one by

$$g(X, Y)(x) = g_x(X_x, Y_x)$$

for X, Y in $\mathcal{X}(M)$ and x in M. To have a converse we would have a R-linear extension $T_x M \underset{E}{\to} \mathcal{X}(M)$ for any x in M. In classical differential geometry, this can be done using local chart and flat function. In synthetic differential geometry we need more structure on M. But we do not enter in that matter now.

A structure as a pointwise Riemannian structure is a good mean to handle dual of $T_x M$. In fact $(T_x M)^*$ is isomorph to $T_x M$ and so is microlinear (and of course Euclidean but the dual $L(E, R)$ of a R-module is always Euclidean).

In the following discussion we consider a Lie object (A, L) with a Riemannian structure g.

Definition 2. *Let (A, L, g) a Lie object with a Riemannian structure. A Riemannian connection ∇ on L is a connection such that for all X, Y and Z in L :*

$$Z(g(X, Y)) = g(\nabla_Z X, Y) + g(X, \nabla_Z Y).$$

The condition just given is a kind of derivation law for g viewed as a (scalar) product. The algebraic basis of Riemannian geometry essentially lies on the existence of Riemannian connection.

Proposition 1. *Let (A, L, g) be a Lie object with a Riemannian structure. There exists one and only one Riemannian connection with null torsion.*

Proof:

a) First we prove the unicity. Let ∇ be a Riemannian connection with null torsion. As the torsion is zero, we have (cf. section 6.2.3, definition 3)

$$\nabla_Z X = [Z, X] + \nabla_X Z.$$

Substituting in the equality of definition 2, we obtain

$$Z(g(X, Y)) = g(\nabla_X Z, Y) + g(X, \nabla_Z Y) + g([Z, X], Y).$$

We then circularly permute the (X, Y, Z) and compute the alternated sum of these expressions. Taking in account the symmetry of g we have

$$2g(X, \nabla_Z Y) = Z(g(X, Y)) - X(g(Y, Z)) + Y(g(Z, X))$$
$$- g([Z, X], Y) + g([X, Y], Z) - g([Y, Z], X).$$

We know g is non-degenerate, thus this formula uniquely characterizes the element $\nabla_Z Y$ in L.

b) We now define $\nabla_Z Y$ by the last formula.

(i) ∇ is an action on L. By derivation and Koszul's axioms :

$$2g(X, \nabla_Z(aY)) = a \cdot Z(g(X, Y)) + Z(a) \cdot g(X, Y)$$
$$- a \cdot X(g(Y, Z)) - X(a)g(Y, Z) + aY(g(Z, X))$$
$$- a \cdot g([Z, X], Y) + a \cdot g([X, Y], Z)$$
$$+ X(a)g(Y, Z) - a \cdot g([Y, Z], X) + Z(a) \cdot g(Y, X)$$
$$= 2g(X, a \cdot \nabla_Z Y) + 2g(X, Z(a) \cdot Y)$$

and thus

$$\nabla_Z(a \cdot Y) = a \cdot \nabla_Z Y + Z(a) \cdot Y.$$

(ii) The action ∇ is a connection.

$$
\begin{aligned}
2g(X, \nabla_{aZ}Y) &= a \cdot Z(g(X,Y)) - X(a)g(Y,Z) \\
&\quad - a \cdot X(g(Y,Z)) + Y(a) \cdot g(Z,X) + a \cdot Y(g(Z,X)) \\
&\quad - a \cdot g([Z,X],Y) + X(a)g(Z,Y) + a \cdot g([X,Y],Z) \\
&\quad - ag([Y,Z],X) - Y(a)g(Z,X) \\
&= a \cdot 2g(X, \nabla_Z Y) = 2g(X, a \cdot \nabla_Z Y).
\end{aligned}
$$

(iii) The connection ∇ is Riemannian. By the definition of $\nabla_Z Y$ and $\nabla_Z X$, we have

$$
\begin{aligned}
2g(X, \nabla_Z Y) &+ 2g(Y, \nabla_Z X) \\
&= Z(g(X,Y)) - X(g(Y,Z)) + Y(g(Z,X)) \\
&\quad - g([Z,X],Y) + g([X,Y],Z) - g([Y,Z],X) \\
&\quad + Z(g(Y,X)) - Y(g(X,Z)) + X(g(Z,Y)) \\
&\quad - g([Z,Y],X) + g([Y,X],Z) - g([X,Z],Y)
\end{aligned}
$$

and, by the symmetry of g and the antisymmetry of the bracket, this gives

$$
2g(X, \nabla_Z Y) + 2g(Y, \nabla_Z X) = 2Z(g(X,Y)).
$$

(iv) The Riemannian connection ∇ is without torsion. In the same way as in (iii) we compute

$$
2g(X, \nabla_Z Y) - 2g(X, \nabla_Y Z)
$$

and we obtain $2g(X, [Z,Y])$ and thus

$$
\nabla_Z Y - \nabla_Y Z = [Z,Y].
$$

∎

Let (A, L, g) a Lie object with a Riemannian structure. Let ∇ the corresponding Riemannian connection without torsion. Let R be the curvature of ∇ viewed as the map $R : L \times L \to \mathcal{L}_A(L)$ given by

$$
R(X,Y)(Z) = \nabla_X(\nabla_Y(Z)) - \nabla_Y(\nabla_X(Z)) - \nabla_{[X,Y]}(Z).
$$

We shall prove, in this general context, the classical properties of symmetry and antisymmetry.

The first antisymmetry is evident.

Proposition 2. *For all X and Y in L,*

$$R(X,Y) = -R(Y,X).$$

∎

Proposition 3. *For all X, Y, Z, V in L, we have the antisymmetry :*

$$g(R(X,Y)(Z), V) + g(R(X,Y)(V), Z) = 0.$$

Proof: By definition of R :

$$g(R(X,Y)(Z), V) + g(R(X,Y)(V), Z)$$
$$= g(\nabla_X \nabla_Y Z, V) - g(\nabla_Y \nabla_X Z, V) - g(\nabla_{[X,Y]} Z, V)$$
$$+ g(\nabla_X \nabla_Y V, Z) - g(\nabla_Y \nabla_X V, Z) - g(\nabla_{[X,Y]} V, Z).$$

As ∇ is Riemannian we have

$$g(\nabla_X \nabla_Y Z, V) = X(g(\nabla_Y Z, V)) - g(\nabla_Y Z, \nabla_X V)$$

and similar equalities obtained by permutation of X and Y and of Z and V. We get, after some simplifications,

$$g(R(X,Y)(Z), V) + g(R(X,Y)(V), Z)$$
$$= X(g(\nabla_Y Z, V)) + X(g(\nabla_Y V, Z))$$
$$- Y(g(\nabla_X Z, V)) - Y(g(\nabla_X V, Z))$$
$$- g(\nabla_{[X,Y]} Z, V) - g(\nabla_{[X,Y]} V, Z).$$

Using six times the definition of the Riemannian connection ∇ we obtain

$$g(R(X,Y)(Z), V) + g(R(X,Y)(V), Z)$$
$$= \frac{1}{2}\Big(X(Y(g(Z,V))) - Y\big(X(g(Z,V))\big) - [X,Y](g(Z,Y))\Big)$$

and this is zero by the Lie-module axiom. ∎

Proposition 4. *For all X, Y, Z, V in L, we have the symmetry :*

$$g(R(X,Y)(Z), V) = g(R(Z,V)(X), Y).$$

Proof: This is just a computation using definition of R, characterization of a Riemannian connection (definition 2), the nullity of the torsion and the formula giving ∇, the Riemannian connection without torsion. ∎

7.1.2   Pre-symplectic and symplectic structures

We can mimic the Riemannian case with skew-symmetric maps in place of symmetric ones. We so obtain a concept of pre-symplectic structure. But the good concept includes a closeness condition.

   However, for clarity, we begin with the three pre-symplectic concepts.

Definition 3.

   1) *Let M be a microlinear object. A pointwise pre-symplectic structure on M is the datum for every $x \in M$ of a map*

$$h_x : T_x M \times T_x M \to R$$

   R-bilinear, antisymmetric and nondegenerate.

   2) *A global pre-symplectic structure on M is a map*

$$h : \mathcal{X}(M) \times \mathcal{X}(M) \to R^M$$

   R^M-bilinear, antisymmetric and nondegenerate.

   3) *Let (A, L) be a Lie object. A pre-symplectic structure on (A, L) is a map*

$$h : L \times L \to A$$

   A-bilinear, antisymmetric and non-degenerate.

   Here are some generalities about A-bilinear non-degenerate forms $h : L \times L \to A$. We consider

$$h_1 : L \to \mathcal{L}_A(L, A) : X \mapsto h_1(X) = h(X, -)$$

$$h_2 : L \to \mathcal{L}_A(L, A) : Y \mapsto h_2(Y) = h(-, Y).$$

We denote by $\check{h}_1$ and $\check{h}_2$ the inverses of h_1 and h_2. Both in the symmetric and antisymmetric case we write $\hat{h}$ for h_1 and $\check{h}$ for $\check{h}_1$.

   The following proposition gives a trivial description of the "bidual" map

$$\iota : L \to \mathcal{L}_A(\mathcal{L}_A(L, A), A)$$

defined by $\iota(X)(\alpha) = \alpha(X)$ for $X \in L$ and $\alpha \in \mathcal{L}_A(L, A)$.

Proposition 5. *We have*

$$\iota = \mathcal{L}_A(\check{h}_2, A) \circ h_1 = \mathcal{L}_A(\check{h}_1, A) \circ h_2.$$

In particular, if h is symmetric (the Riemannian case where $h_1 = h_2 = \hat{h}$),

$$\iota = \mathcal{L}_A(\check{h}, A) \circ \hat{h}$$

and, if h is antisymmetric (the pre-symplectic case where $\hat{h}_2 = -\hat{h}_1 = -\hat{h}$)

$$-\iota = \mathcal{L}_A(\check{h}, A) \circ \hat{h}.$$

Proof: is a trivial verification.   ∎

In both the symmetric and the antisymmetric cases, the isomorphism $\hat{h} : L \to \mathcal{L}_A(L, A)$ permits to define a Lie algebra structure on $\mathcal{L}_A(L, A)$ by

$$[\alpha, \beta] = \hat{h}([\check{h}(\alpha), \check{h}(\beta)]).$$

We also have an action of $\mathcal{L}_A(L, A)$ on A by

$$\alpha(a) = \check{h}(\alpha)(a)$$

and so we immediately have a kind of dual of a Lie object :

Proposition 6. $(A, \mathcal{L}_A(L, A))$ *is a Lie object.*   ∎

We can extend $h : L \times L \to A$ to $\mathcal{L}_A(L, A)$, defining

$$\tilde{h} : \mathcal{L}_A(L, A) \times \mathcal{L}_A(L, A) \to A$$

by $\tilde{h}(\alpha, \beta) = h(\check{h}(\alpha), \check{h}(\beta))$. The fact that h is non-degenerate implies the same for $\tilde{h}$. $\tilde{h}$ is symmetric (resp. antisymmetric) if and only if so is h.

Now we fix some classical points of terminology.

Definition 4. *If $h : L \times L \to A$ is A-bilinear and non-degenerate, we define the h-gradient map*

$$G(= G^{(h)}) : A \to L$$

by

$$G^{(h)}(a) = \check{h}(da).$$

Proposition 7. *The h-gradient G is R-linear and is a derivation :*

$$G(ab) = aG(b) + bG(a).$$

Proof: By definition of G :

$$
\begin{aligned}
h(G(ab), X) &= d(ab)(X) \\
&= a \cdot db(X) + b \cdot da(X) \\
&= ah(Gb, X) + bh(Ga, X) \\
&= h(aGb + bGa, X)
\end{aligned}
$$

and so the result as h is non degenerate. ∎

In the symmetric case, one says $G(a)$ is the gradient of a. If h is antisymmetric one says $G^{(h)}(a)$ is a Hamiltonian field of energy a. We often denote it by H_a.

Proposition 8. *For α in $\mathcal{L}_A(L, A)$ and a in A,*

$$\check{h}(\alpha)(a) = \pm\alpha(G(a))$$

(the sign is $+$ if h is symmetric and $-$ if h is antisymmetric).

Proof: By proposition 4

$$\pm\iota = \mathcal{L}_A(\check{h}, A) \circ \hat{h}.$$

And thus for all a and α

$$
\begin{aligned}
\pm\alpha(Ga) &= \pm\iota(G(a))(\alpha) \\
&= da(\check{h}(a)) \\
&= \check{h}(\alpha)(a).
\end{aligned}
$$

∎

We can mimic the construction of the Riemannian connection in the case of a pre-symplectic structure.

Definition 5. *Let (A, L) be a Lie object with a pre-symplectic structure h. For Y, Z in L, we define $\nabla_Z Y$ by*

$$
\begin{aligned}
2h(X, \nabla_Z Y) = {}&Z(h(X, Y)) + X(h(Y, Z)) - Y(h(Z, Y)) \\
&- h([Z, X], Y) - h([X, Y], Z) + h([Y, Z], X)
\end{aligned}
$$

for every X in L.

Proposition 9. *Definition 5 gives a connection on* (A, L).

Proof:

(i) $\nabla_Z X$ is an action. By the derivation and Koszul's axioms :

$$2h(X, \nabla_Z aY) = aZ(h(X, Y)) + Z(a)h(X, Y)$$
$$+ aX(h(Y, Z)) + X(a)h(Y, Z) - aY(h(Z, X))$$
$$- ah([Z, X], Y) - ah([X, Y], Z)$$
$$- X(a)h(Y, Z) + ah([Y, Z], X) - Z(a)h(Y, X)$$
$$= 2ah(X, \nabla_Z Y) + Z(a)h(X, Y) - Z(a)h(Y, X)$$

and thus, as h is antisymmetric,

$$h(X, \nabla_Z aY) = ah(X, \nabla_Z Y) + h(X, Z(a) \cdot Y).$$

So $\nabla_Z aY = a\nabla_Z Y + Z(a)Y$ and ∇ is an action.

(ii) The action ∇ is a connection. We have

$$2h(X, \nabla_{aZ} Y) = aZ(h(X, Y)) + aX(h(Y, Z))$$
$$+ X(a)h(Y, Z) - aY(h(Z, X)) - Y(a)h(Z, X)$$
$$- ah([Z, X], Y) + X(a)h(Z, Y) - ah([X, Y], Z)$$
$$+ ah([Y, Z], X) + Y(a)h(Z, X)$$
$$= 2ah(X, \nabla_Z Y)$$

and $\nabla_{aZ} Y = a\nabla_Z Y$.

∎

Some calculations give results with some analogies with the case of Riemannian connections. For example we have a kind of quasi-derivation formula :

$$Z(h(X, Y)) = h(\nabla_Z X, Y) + h(X, \nabla_Z Y) + h([X, Y], Z)$$

or

$$h(\nabla_Z X - [X, Z], Y) + h(\nabla_Y X - [X, Y], Z) = 0.$$

But the connection just described is not without torsion and is not so much interesting as in the Riemannian case.

Now we come to the description of symplectic structure. Of course, a pre-symplectic structure $h : L \times L \to A$ is a 2-form on (A, L) in the sense of definition 5 in 6.1.2. So the following definition makes sense.

Definition 6. *Let (A, L) be a Lie object. A symplectic structure on (A, L) is a 2-form $h : L \times L \to A$ non-degenerate and closed. We say (A, L, h) is a symplectic object.*

We remind that h closed means $dh = 0$, i.e. $\forall X, Y, Z \in L$,

$$X(h(Y, Z)) - Y(h(X, Z)) + Z(h(X, Y))$$
$$- h([X, Y], Z) + h([X, Z], Y) - h([Y, Z], X) = 0.$$

Definition 7. *Let (A, L, h) be a symplectic object. An element X in L is locally Hamiltonian if the 1-form $i_X h$ $(= \hat{h}(X))$ is closed. We say X is Hamiltonian if $i_X h$ is exact. We note $LH(A, L, h)$ (resp. $H(A, L, h)$) the R-submodule of L consisting of the elements of L locally Hamiltonian (resp. Hamiltonian).*

Proposition 10. *$LH(A, L, h)$ is a Lie subalgebra of L, $H(A, L, h)$ is an ideal in $LH(A, L, h)$. Furthermore*

$$[LH(A, L, h), LH(A, L, h)] \subseteq H(A, L, h),$$

or more explicitely

$$i_{[X_1, X_2]} h = -d(h(X_1, X_2))$$

for X_1, X_2 in $LH(A, L, h)$.

Proof: It suffices to verify the last affirmation. This comes from the Cartan's identities (6.1.2, proposition 6) :

$$i_{[X_1, X_2]} h = L_{X_1} i_{X_2} h - i_{X_2} L_{X_1} h$$
$$= i_{X_1} d i_{X_2} h + d i_{X_1} i_{X_2} h$$
$$\quad - i_{X_2} i_{X_1} dh + i_{X_2} d i_{X_1} h$$
$$= d i_{X_1} i_{X_2} h$$
$$= d(h(X_2, X_1))$$

(because $i_{X_1} h, i_{X_2} h$ and h are closed). ∎

Proposition 11. *(A, L, h) is a symplectic object if and only if so is $(A, \mathcal{L}_A(L, A), \tilde{h})$.*

Proof: Obviously

$$(\tilde{dh})(\alpha, \beta, \gamma) = (dh)(\check{h}(\alpha), \check{h}(\beta), \check{h}(\gamma))$$

and so $\tilde{h}$ is closed if and only if h is closed. ∎

Corresponding to proposition 9, we have

Proposition 12. *Let* (A, L, h) *be a symplectic object. If* α *and* β *are closed 1-forms, then* $[\alpha, \beta]$ *is an exact 1-form. More precisely,*

$$[\alpha, \beta] = -d(\tilde{h}(\alpha, \beta)).$$

∎

In the symplectic situation we rename the h-gradient $H : A \to L$ as *Hamilton operator.* We have $H(a) = \check{h}(da)$. Thus

$$h(H(a), X) = da(X) = X(a).$$

We have thus the following proposition.

Proposition 13. *For all* $a, b \in A$,

$$h(H(a), H(b)) = \tilde{h}(da, db) = -H_a(b) = H_b(a).$$

∎

Definition 8. *The Poisson bracket*

$$\{-, -\} : A \times A \to A$$

is given by $\{a, b\} = h(H(a), H(b))$.

It is also possible to see A as operating on itself by the A-linear action

$$a(b) = \{a, b\}.$$

The following proposition sums up the algebraic properties of the Poisson bracket.

Proposition 14. *Let (A, L, h) be a symplectic object.*

1) The Poisson bracket induces on A a Lie algebra structure ;

2) (A, A) is a Lie object (for the action $a(b) = \{a, b\}$) ;

3) In particular
$$\{a, bc\} = b\{a, c\} + \{a, b\}c$$

4) $-H : A \to L$ is an homomorphism of Lie algebra :

$$H(\{a, b\}) = -[H(a), H(b)].$$

Proof: First we prove (4) :

$$H(\{a, b\}) = \check{h}(d(\{a, b\}))$$

(by definition 4)

$$= \check{h}\Big(d(\tilde{h}(da, db))\Big)$$

(proposition 13)

$$= \check{h}(-[da, db])$$

(proposition 12)

$$\begin{aligned} &= -[\check{h}(da), \check{h}(db)] \\ &= -[H(a), H(b)]. \end{aligned}$$

As for (1), R-bilinearity and antisymmetry of Poisson bracket beeing trivial, only we must verify Jacobi identity. By proposition 12, we have

$$\begin{aligned} \{a, \{b, c\}\} &= -H_a(\{b, c\}) = H_a(H_b(c)) \\ \{b, \{c, a\}\} &= -H_b(\{c, a\}) = -H_b(H_a(c)) \\ \{c, \{a, b\}\} &= H_{\{a,b\}}(c). \end{aligned}$$

But by (4) just proved, this is

$$\begin{aligned} \{c, \{a, b\}\} &= -[H(a), H(b)](c) \\ &= -H(a)(H(b)(c)) + H(b)(H(a)(c)) \end{aligned}$$

and proves Jacobi identity.

Now we verify that (A, A) is a Lie object. The Lie-module axiom reduces to

$$\{\{a, b\}, c\} = \{a, \{b, c\}\} - \{b, \{a, c\}\}$$

which comes from Jacobi identity. Both derivation and Koszul's axioms reduce to (3) :

$$\{a, bc\} = b\{a, c\} + \{a, b\}c$$

which comes from proposition 7. ∎

Here we insert a note on Poisson structure.

Definition 9. *Let A be a commutative R-algebra. A Poisson structure P on A is a map*

$$P : A \times A \to A$$

which is a biderivation and provides A with a Lie algebra structure. We note $\{a, b\}$ for $P(a, b)$.

If (A, L) is a Lie object and P a Poisson structure on A, we say (A, L, P) is a *Poisson object.* Of course, if (A, L, h) is a symplectic object, the Poisson bracket (definition 8) is a Poisson structure on A (proposition 14).

To give another example, also classical, we explain the Kostant-Souriau case. Let $\mathcal{G}$ be a reflexive, Euclidean, microlinear Lie algebra. By reflexivity we mean the canonical map $\iota : \mathcal{G} \to \mathcal{G}^{**}$ is bijective. We consider $A = R^{\mathcal{G}^*}$. For $f, g \in A$ and $\alpha \in \mathcal{G}^*$, $df(\alpha)$ and $dg(\alpha)$ are in $\mathcal{G}^{**}$. Then we put

$$\{f, g\}(\alpha) = \alpha\left([\iota^{-1}(df(\alpha)), \iota^{-1}(dg(\alpha))]\right)$$

and we obtain a Poisson structure on A.

If M is a microlinear object, a Poisson structure P on M is a Poisson structure on $A = R^M$. It is the case of Kostant-Souriau for $M = \mathcal{G}^*$.

Let $\mathrm{Der.}(A)$ and $\mathrm{Der}_P(A)$ the Lie algebras of derivations of A for respectively the multiplicative structure on A and the Lie structure (given by P) on A. Trivially we have

Proposition 15. *The map*

$$A \to A^A : f \mapsto \{f, -\}$$

is a Lie algebra homomorphism from (A, P) to $\mathrm{Der.}(A) \cap \mathrm{Der}_P(A)$. ∎

Now we make some easy algebraic observations in the case of reflexive Lie objects, i.e. Lie objects of the form $(A, \text{Der } A)$. We know $(R^M, \mathcal{X}(M))$ is a such reflexive Lie object when M is a reflexive microlinear object (cfr. 3.3.2).

We denote by X_f the derivation associated to f by $X_f(g) = \{f, g\}$; X_f is in Der A.

Definition 10. *Let $(A, \text{Der } A)$ be a reflexive Lie object and P a Poisson structure on A. The Poisson tensor is the A-linear map*

$$\Lambda : \mathcal{L}_A(L, A) \to L$$

given by

$$\Lambda(\alpha)(f) = \alpha(X_f)$$

(where $L = \text{Der} A$).

Proposition 16. *Let $(A, \text{Der } A = L)$ be a reflexive Lie object. The following data are equivalent*

1) *A poisson structure P on A ;*

2) *A R-linear map $X : A \to L = \text{Der } A$ such that :*

   (a) *X is a derivation from A to the A-module L, i.e. $X_{fg} = f \cdot X_g + g \cdot X_f$.*
   (b) *X is antisymmetric, i.e.*

   $$X_f(g) = -X_g(f).$$

   (c) *For all f, g, h in A,*

   $$X_f X_g h + X_g X_h f + X_h X_f g = 0.$$

3) *A A-linear map $\Lambda : \mathcal{L}_A(L, A) \to L$ such that*

   (a) *For all f and g in A*

   $$\Lambda(df)(g) = -\Lambda(dg)(f).$$

(b) For all f, g and h in A

$$\Lambda(df)(\Lambda(dg)(h)) + \Lambda(dg)(\Lambda(dh)(f))$$

$$+\Lambda(dh)(\Lambda(df)(g)) = 0.$$

Proof: Trivial computations with the following constructions :
a) Define X from P by $X_f(g) = \{f, g\}$;
b) Define Λ from X by $\Lambda(\alpha)(f) = \alpha(X_f)$;
c) Define P from Λ by $\{f, g\} = \Lambda(dg)(f)$.   ∎

Definition 11. *Let $(A, \text{Der } A, P)$ be a reflexive Poisson object*

a) $X \in \text{Der } A$ *is a Poisson element if it is also in $\text{Der } A_P$;*

b) *A 1-form α is a Poisson form if $\Lambda(\alpha)$ is a Poisson element.*

Proposition 17. *Let $(A, \text{Der } A, P)$ be a reflexive Poisson object. A 1-form α is a Poisson form if and only if*

$$d\alpha(X_f, X_g) = 0$$

for all f and g in A.

Proof: α is Poisson if and only if $\forall f, g \in A$

$$\Lambda(\alpha)(\{f, g\}) = \{\Lambda(\alpha)f, g\} + \{f, \Lambda(\alpha)g\}$$

iff

$$\alpha(X_{\{f,g\}}) = \{\alpha(X_f), g\} + \{f, \alpha(X_g)\}$$

iff

$$\alpha([X_f, X_g]) = X_g(\alpha(X_f)) - X_f(\alpha(X_g))$$

iff

$$d\alpha(X_f, X_g) = 0.$$

∎

7.1.3   Complex situations

In this subsection we suppose a Lie object (A, L) to be fixed.

Definition 12. *An* almost complex structure *on (A, L) is an A-linear endomorphism J on L such that $J^2 = -1$. We say (A, L, J) is an almost* complex object.

We consider the algebra $A[i]$ with $i^2 = -1$ and L becomes an $A[i]$-module by :

$$(a + ib)X = aX + bJ(X).$$

Definition 13. *An element X in L is* J-adapted *if $J \circ \mathrm{ad}_X = \mathrm{ad}_X \circ J$, i.e. for all Y in L, $J([X, Y]) = [X, JY]$.*

Proposition 18. *The set of J-adapted elements in L is a R-Lie subalgebra of L.*

Proof: Let X and Y be J-adapted. By Jacobi's identity, for all Z in L,

$$J([[X, Y], Z]) = -J([[Y, Z], X]) - J([[Z, X], Y])$$

and thus, as X and Y are J-adapted

$$\begin{aligned} J([[X, Y], Z]) &= -[[Y, JZ], X] - [[JZ, X], Y] \\ &= [[X, Y], JZ] \end{aligned}$$

and $[X, Y]$ is J-adapted.   ∎

The Lie algebra of J-adapted elements is denoted by L_J.

Definition 14. *Let (A, B, J) an almost complex object. The* torsion *of J is the map*

$$N : L \times L \to L$$

given by

$$N(X, Y) = [JX, JY] - [X, Y] - J([X, JY] + [JX, Y]).$$

Proposition 19. *If X and Y are J-adapted, $N(X, Y) = [JX, JY] + [X, Y]$.*

Proof: Use the definition of N, the fact that X and Y are J-adapted and $J^2 = -1$.   ∎

Definition 15. *Let (A, L, J) be an almost complex object. We say (A, L, J) is a complex object or J is a complex structure, if the torsion is zero.*

Proposition 20. *Let J be a complex structure on (A, L) ; if X is J-adapted then JX is J-adapted. So J becomes an A-linear map from L_J into itself.*

Proof: As X is J-adpated the torsion $N(X, Y)$ becomes, for every Y in L,
$$[JX, JY] - J([JX, Y]).$$
So the nullity of $N(X, Y)$ gives that JX is J-adapted.   ∎

Let C be $R[i]$. Of course L is a C-module by
$$(\alpha + i\beta)X = \alpha X + \beta JX.$$

If J is only an almost complex structure, the Lie bracket may not to be C-bilinear. But we have

Proposition 21. *If J is a complex structure, L_J is a C-Lie algebra.*

Proof: It suffices to see
$$[X, iY] = [X, JY] = J([X, Y]) = i[X, Y].$$

    ∎

Let $C(L)$ be the complexification of L. Our aim is to describe holomorph elements in $C(L)$.

First we extend J to $C(L)$:
$$J(X + iY) = JX + iJY$$
and J is $C(A)$-linear. We can describe $C(L)$ as the direct sum $L^1 \oplus L^2$ with
$$
\begin{aligned}
L^1 &= \{Z \in C(L) | JZ = iZ\} \\
L^2 &= \{Z \in C(L) | JZ = -iZ\}.
\end{aligned}
$$
Of course $X + iY \in L^1$ iff $Y = -JX$ and $X + iY \in L^2$ iff $Y = JX$.

Proposition 22. *If J is a complex structure, L^1 and L^2 are C-Lie algebras.*

Proof: If $Z_1 = X_1 + iY_1 \in L^1$ and $Z_2 = X_2 + iY_2 \in L^1$ we have $Y_1 = -JX_1$ and $Y_2 = -JX_2$. So,

$$[Z_1, Z_2] = [X_1, X_2] - [JX_1, JX_2] - i([X_1, JX_2] + [JX_1, X_2]).$$

The torsion is zero, so

$$[JX_1, JX_2] - [X_1, X_2] = J([X_1, JX_2] + [JX_1, X_2])$$

and thus,

$$[Z_1, Z_2] = -J([X_1, JX_2] + [JX_1, X_2]) - i([X_1, JX_2] + [JX_1, X_2])$$

and $[Z_1, Z_2] \in L^1$. The proof is similar for L^2. ∎

Proposition 23. *Let always J be a complex structure. The map*

$$\varphi : L_J \to L^1 : X \to \frac{1}{2}(X - iJX)$$

is an injective homomorphism of C-Lie algebras.

Proof: We only must prove that $\varphi([X, Y]) = [\varphi X, \varphi Y]$. We have

$$\varphi([X, Y]) = \frac{1}{2}([X, Y] - iJ([X, Y]))$$

and

$$[\varphi X, \varphi Y] = \frac{1}{4}([X, Y] - [JX, JY] - i([X, JY] + [JX, Y])).$$

As X and Y are in L_J, $[X, JY] = [JX, Y] = J([X, Y])$ and $[JX, JY] = -[X, Y]$. Then

$$[\varphi X, \varphi Y] = \frac{1}{2}([X, Y] - iJ([X, Y])) = \varphi([X, Y]).$$

∎

Then we can do the following definition.

Definition 16. *Z in $C(L)$ is holomorph if Z is in the image of φ.*

Now we turn to almost complex structure on A-modules. Let E be an A-module. An almost complex structure on E is an A-linear endomorphism J of E such that $J^2 = -\text{id}_E$. Then E becomes an $A[i]$-module by $i \cdot v = J(v)$ for $v \in E$. Let (E, J) and (E^1, J^1) be two almost complex A-modules. A map $U : E \to E^1$ is $A[i]$-linear if and only if, it is A-linear and $U \circ J = J^1 \circ U$. On $L_{A[i]}(E, E^1)$ we define an almost complex structure $\bar{J}$ by

$$\bar{J}(U) = J^1 \circ U = U \circ J.$$

We now have almost complex connections.

Definition 17. *Let (E, J) an almost complex A-module. A almost complex connection $\nabla : L \times E \to E$ is a connection such that for all X in L, $\nabla_X \circ J = J \circ \nabla_X$.*

Proposition 24. *To a connection ∇ on E, corresponds an almost complex connection $\hat{\nabla}$ defined by*

$$2\hat{\nabla}(X, v) = \nabla(X, v) - J(\nabla(X, JV)).$$

Proof: That $\hat{\nabla}$ is a connection is trivial. We show $\hat{\nabla}$ is almost complex :

$$
\begin{aligned}
2\hat{\nabla}(X, Jv) &= \nabla(X, Jv) - J(\nabla(X, J^2 v)) \\
&= -J^2(\nabla(X, Jv)) + J(\nabla(X, v)) \\
&= 2J(\hat{\nabla}(X, v)).
\end{aligned}
$$

∎

If (A, L, J) is an almost complex object and ∇ is an almost complex connection on L, we must see the relationship of the torsion of J with the torsion of ∇.

Proposition 25. *If ∇ is an almost complex connection on L, we have*

$$N(X, Y) = T(X, Y) + J(T(JX, Y)) + J(T(X, JY)) - T(JX, JY).$$

Proof: This results from a trivial computation using the definition of T and $\nabla_X \circ J = J \circ \nabla_X$.

∎

7.1.4   Hermitian and Kaelherian objects

Let (A, L, J) an almost complex object and h a Riemannian structure $h : L \times L \to A$.

Definition 18. *We say that h is an Hermitian structure if, for any X and Y in L*

$$h(JX, JY) = h(X, Y).$$

If (A, L, J) is almost complex (resp. complex) and h is an Hermitian structure, we say that (A, L, J, h) is an almost Hermitian object (resp. an Hermitian object).

On Riemannian almost complex objects there always exists an Hermitian structure.

Proposition 26. *Let (A, L, J) an almost complex object and g a Riemannian structure on (A, L). The map $h : L \times L \to A$ given by*

$$h(X, Y) = g(X, Y) + g(JX, JY)$$

is an Hermitian structure.   ■

Instead of the symmetry condition of Riemannian structure, we can define Hermitian structures by 2-form.

Proposition 27. *Let (A, L, J) be an almost complex object. There is a bijection between the set of Hermitian structure on (A, L, J) and the set of non-degenerate and J-invariant 2-forms.*

Proof: To each map $h : L \times L \to A$ we associate $\omega : L \times L \to A$ by $\omega(X, Y) = h(X, JY)$ and to $\omega : L \times L \to A$ we associate $h : L \times L \to A$ by $h(X, Y) = \omega(JX, Y)$. For the two directions we observe trivially : h is A-bilinear iff ω is A-bilinear, h is non-degenerate iff ω is non-degenerate, h is compatible with J iff ω is compatible with J. If h and ω are compatible with J, h is symmetric iff ω is antisymmetric. For example

$$
\begin{aligned}
\omega(Y, X) &= h(Y, JX) \\
&= h(JY, JJX) \\
&= -h(JY, X) \\
&= -\omega(X, Y).
\end{aligned}
$$

For J-compatible h and ω, the described correspondence is bijective. ■

We point out the construction of a sesquilinear Hermitian form associated to an Hermitian structure.

Proposition 28. *Let (A, L, J, h) be an almost Hermitian object. We have, on the $A[i]$-module L, the sesquilinear Hermitian form with values in $A[i]$, given by*

$$\varphi(X, Y) = h(X, Y) + i\omega(X, Y).$$

Proof: Trivial computations. For example

$$
\begin{aligned}
\varphi(JX, Y) &= h(JX, Y) + i\omega(JX, Y) \\
&= -\omega(X, Y) + ih(X, Y) \\
&= i\varphi(X, Y)
\end{aligned}
$$

and, in the same way, $\varphi(X, JY) = -i\varphi(X, Y)$. ■

If ∇ is the Riemannian connection without torsion associated to (A, L, h), ∇ is not, in general, J-compatible (i.e. almost complex in the sense of definition 17). The difference between $\nabla_X JY$ and $J\nabla_X Y$ depends on N and $d\omega$.

Proposition 29. *If (A, L, J, h) is an almost Hermitian object with ω the associated 2-form, N the torsion of J, ∇ the Riemannian connection without torsion associated to (A, L, h). We have*

$$
\begin{aligned}
2h(\nabla_X JY - J\nabla_X Y, Z) &= d\omega(X, JY, JZ) \\
&- d\omega(X, Y, Z) + h(N(Y, Z), JX).
\end{aligned}
$$

Proof: The lefthandside of our thesis is

$$
\begin{aligned}
2h(\nabla_X JY &- J\nabla_X Y, Z) \\
&= 2h(\nabla_X JY, Z) + 2h(\nabla_X Y, JZ) \\
&= X(h(Z, JY)) - Z(h(JY, X)) + (JY)(h(X, Z)) \\
&- h([X, Z], JY) + h([Z, JY], X) - h([JY, X], Z) \\
&+ X(h(JZ, Y)) - (JZ)(h(Y, X)) + Y(h(X, JZ)) \\
&- h([X, JZ], Y) + h([JZ, Y], X) - h([Y, X], JZ).
\end{aligned}
$$

We know that

$$dw(A, B, C) = A(\omega(B, C)) - B(\omega(A, C)) + C(\omega(A, B))$$
$$- \omega([A, B], C) + \omega([A, C], B) - \omega([B, C], A)$$
$$N(A, B) = [JA, JB] - [A, B] - J([A, JB] + [JA, B])$$
$$\omega(A, B) = h(A, JB) = -h(JA, B).$$

Let righthandside of the thesis is

$$X(\omega(JY, JZ)) - (JY)(\omega(X, JZ)) + (JZ)(\omega(X, JY))$$
$$- \omega([X, JY], JZ) + \omega([X, JZ], JY) - \omega([JY, JZ], X)$$
$$- X(\omega(Y, Z)) + Y(\omega(X, Z)) - Z(\omega(X, Y))$$
$$+ \omega([X, Y], Z) - \omega([X, Z], Y) + \omega([Y, Z], X)$$
$$+ h([JY, JZ], JX) - h([Y, Z], JX)$$
$$- h(J([Y, JZ]), JX) - h(J([JY, Z]), JX)$$
$$= X(h(Y, JZ)) + (JY)(h(X, Z)) - (JZ)(h(X, Y))$$
$$+ h([X, JY], Z) - h([X, JZ], Y) - h([JY, JZ], JX)$$
$$- X(h(Y, JZ)) + Y(h(X, JZ)) - Z(h(X, JY))$$
$$+ h([X, Y], JZ) - h([X, Z], JY) + h([Y, Z], JX)$$
$$+ h([JY, JZ], JX) - h([Y, Z], JX)$$
$$- h(J([Y, JZ]), JX) - h(J([JY, Z]), JX).$$

By the J-invariance, we have the results.   ∎

From the purely algebraic point of view, the following definition will appears as natural.

Definition 19. Let (A, L, J, h) an almost Hermitian object and ω the associated 2-form. (A, L, J, h) is a Kaelherian object if $N = 0$ and $dw = 0$.

Proposition 30. Let (A, L, J, h) an almost Hermitian object, ω the associated 2-form and ∇ the Riemannian connection without torsion associated to (A, L, h). The following are equivalent

(i) (A, L, J, h) is a Kaelherian object ;

(ii) (A, L, J) *is a complex object and* (A, L, ω) *is a symplectic object ;*

(iii) ∇ *is almost complex.*

Proof: That (i) $\Longleftrightarrow$ (ii) is just matter of definition. If $N = 0$ and $d\omega = 0$, by proposition 29 we have

$$h(\nabla_X JY - J\nabla_X Y, Z) = 0.$$

And, as h is non degenerated, ∇ is almost complex. It remains to prove that (iii) $\Rightarrow$ (i). We suppose ∇ almost complex. As the torsion T of ∇ is zero, we have $N = 0$ by proposition 25. We must prove $d\omega = 0$. By the definition of a Riemannian connection (definition 2), we have

$$Z(h(X, JY) - h(\nabla_Z X, JY) - h(X, \nabla_Z JY) = 0$$

and thus, as ∇ is almost complex

$$Z(h(X, JY) - h(\nabla_Z X, JY) - h(X, J\nabla_Z Y) = 0$$

or

$$Z(\omega(X, Y)) - \omega(\nabla_Z X, Y) - \omega(X, \nabla_Z Y) = 0$$

or, by the notation of 6.2.3 (proposition 8)

$$(\nabla_X \omega)(Y, Z) = 0.$$

We have thus, by this proposition 8 of 6.2.3, $d\omega = 0$.   ∎

7.2   Lie algebras of Lie groups

Later, for some global examples of microlinear groups, we shall need the general concept of Lie derivative.

First, let M be a microlinear object. Let $u \in \text{Iso}(M)$ a bijective map from M to M. We will show how u operates on some elements associated with M.

a) If $f : M \to R$ is a map, we define $u^*(f)$ as $f \circ u$. So we obtain

$$u^* : R^M \to R^M.$$

It is an automorphism of R-algebras. Perhaps it is interesting to quote that

$$u^* : R - \mathrm{alg}(R^M, R) \to R - \mathrm{alg}(R^M, R)$$

defined by $u^*(A) = A \circ u^*$, make the following diagram commutative

$$
\begin{array}{ccc}
M & \xrightarrow{ev} & R - \mathrm{alg}(R^M, R) \\
\downarrow{u} & & \downarrow{u^*} \\
M & \xrightarrow{ev} & R - \mathrm{alg}(R^M, R)
\end{array}
$$

b) Also we define

$$u^* : \mathcal{X}(M) \to \mathcal{X}(M)$$

by $u^*(X)(d) = u^{-1} \circ X(d) \circ u$ for any X in $\mathcal{X}(M)$ and d in D. In other words,

$$u^*(X)(x, d) = u^{-1}(X(u(x), d)).$$

Proposition 1. *The map* $u^* : \mathcal{X}(M) \to \mathcal{X}(M)$ *is Lie algebras automorphism.*

Proof: First

$$
\begin{aligned}
(\alpha \cdot u^*(X))(x, d) &= u^*(X)(x, \alpha \cdot d) \\
&= u^{-1}(X(u(x), \alpha \cdot d)) \\
&= u^{-1}((\alpha X)(u(x), d)) \\
&= u^*(\alpha X)(x, d),
\end{aligned}
$$

and so u^* is homogenous and thus R-linear.

That u^* is a Lie algebra homomorphism is the following trivial computation

$$[u^*X, u^*Y](x, d_1 \cdot d_2)$$
$$= \left((u^*Y)_{-d_2} \circ (u^*X)_{-d_1} \circ (u^*Y)_{d_2} \circ (u^*X)_{d_1}\right)(x)$$
$$= u^{-1}(Y(X(Y(X(u(x), d_1), d_2), -d_1), -d_2))$$
$$= u^*([X, Y])(x, d_1 \cdot d_2).$$

As obviously $(u^{-1})^* = (u^*)^{-1}$, the proof is complete. ∎

Also we can consider the R^M-module structure on $\mathcal{X}(M)$ defined by

$$(f \cdot X)(x, d) = X(x, f(x) \cdot d).$$

Proposition 2. $u^* : \mathcal{X}(M) \to \mathcal{X}(M)$ is R^M-compatible in that sense :

$$u^*(f \cdot X) = u^*(f) \cdot u^*(X).$$

Proof: Just the computation

$$
\begin{aligned}
u^*(f \cdot X)(x, d) &= u^{-1}((f \cdot X)(u(x), d)) \\
&= u^{-1}(X(u(x), f(u(x)) \cdot d) \\
&= (u^*f \cdot u^*X)(x, d).
\end{aligned}
$$

∎

c) We can extend this pull-back map u^* to other maps. Consider

$$\varphi : \mathcal{X}(M) \times \ldots \times \mathcal{X}(M) \to R^M$$

R-p-linear. We define $u^*(\varphi)$ by :

$$u^*(\varphi)(X_1, \ldots, X_p) = \varphi((u^{-1})^*(X_1), \ldots, (u^{-1})^*(X_p)) \circ u.$$

It is clear that φ^* is an automorphism of R-algebra of $\mathcal{L}_R(\mathcal{X}(M)^M, R^M)$ (the R-algebra of R-p-linear maps from $\mathcal{X}(M)^p$ into R^M). The most important cases are the cases of symmetric and of antisymmetric forms. Trivially, u^* respects this property of symmetry or antisymmetry.

d) Finally we can consider the set $\mathcal{L}_R(\mathcal{X}(M)^p, \mathcal{X}(M))$ of p-linear maps. We extend u^* by

$$u^*(\Phi)(X_1, \ldots, X_p)(x, d) = u^{-1}(\Phi((u^{-1})^* X_1, \ldots, (u^{-1})^* X_p)(u(x), d))$$

for Φ in $\mathcal{L}_R(\mathcal{X}(M)^p, \mathcal{X}(M))$. As $\mathcal{X}(M)$ is a Lie algebra we have a pointwise defined structure of Lie algebra on $\mathcal{L}_R(\mathcal{X}(M)^p, \mathcal{X}(M))$. It is a simple exercise to see that u^* is a Lie algebra automorphism.

Now, we are ready to describe Lie derivatives. Let M be a microlinear object and X a vector field on M. For all d in D, X_d is bijective with inverse X_{-d}. We can thus study $(X_d)^*(\varphi)$ for different kind of functions φ. For example, for $f \in R^M$, $Y \in \mathcal{X}(M)$, $\varphi \in \mathcal{L}_R(\mathcal{X}(M)^p, R^M)$ or $\Phi \in \mathcal{L}_R(\mathcal{X}(M)^p, \mathcal{X}(M))$, we have $(X_d)^*(f)$, $(X_d)^*(Y)$, $(X_d)^*(\varphi)$ or $(X_d)^*(\Phi)$. In each case $(X_d)^*$ acts on a Euclidean R-module, so we can consider the difference

$$(X_d)^* \varphi - \varphi.$$

But $X_0 = \mathrm{id}_M$, so $(X_0)^*$ is an identity and $(X_0)^*(\varphi) - \varphi$ is zero. By the Euclidean property, the difference can be uniquely written as

$$(X_d)^*(\varphi) - \varphi = d \cdot \psi$$

where ψ is in the same R-module as φ.

Definition 1. For φ in R^M, $\mathcal{X}(M)$, $\mathcal{L}_R(\mathcal{X}(M)^p, R^M)$ or $\mathcal{L}_R(\mathcal{X}(M)^p, \mathcal{X}(M))$, the Lie derivative of φ in the direction of X, denoted $L_X\varphi$, is the element characterized by

$$(X_d)^*(\varphi) - \varphi = d \cdot L_X\varphi$$

for all d in D.

a) If $f \in R^M$, $L_X f$ is given by

$$f \circ X_d - f = d \cdot L_X f$$

and then we meet again the concept of Lie derivative given in 3.3.1 definition 1.

b) If $Y \in \mathcal{X}(M)$ we have the following result.

Proposition 3. *For all X, Y in $\mathcal{X}(M)$,*

$$L_X(Y) = [X, Y].$$

Proof: Let d in D and (d_1, d_1') in $D(2)$. We put

$$l_d(d_1, d_1') = Y_{-d_1} \circ X_{-d} \circ Y_{d_1'} \circ X_d.$$

We have

$$
\begin{aligned}
l_d(d_1, 0) &= Y_{-d_1} = (-Y)_{d_1} \\
l_d(0, d_1') &= X_{-d} \circ Y_{d_1'} \circ X_d \\
&= (X_d)^*(Y)_{d_1'}.
\end{aligned}
$$

Then $l_d : D(2) \to M^M$ is the map such that his restriction to the diagonal is the sum of $(-Y)$ and $(X_d)^*Y$:

$$
\begin{aligned}
l_d(d_1, d_1) &= ((X_d)^*(Y) - Y)_{d_1} \\
&= (d \cdot L_X(Y))_{d_1}.
\end{aligned}
$$

So, for all d in D and d_1 in D,

$$
\begin{aligned}
(d \cdot L_X(Y))_{d_1} &= Y_{-d_1} \circ X_{-d} \circ Y_{d_1} \circ X_d \\
&= [X, Y]_{d \cdot d_1} \\
&= (d \cdot [X, Y])_{d_1}.
\end{aligned}
$$

And $L_X(Y) = [X, Y]$ as we wish. ∎

The two examples just give, as well the general definition $(X_d)^*(\varphi) - \varphi = d \cdot L_X\varphi$, show that we have a differential concept very near of the usual concept of derivative.

c) But algebraically L_X is also a kind of derivation in a tensorial meaning. More explicitely we have the following.

Proposition 4. *Let φ in $\mathcal{L}_R(\mathcal{X}(M)^p, R^M)$, $X, Y_1, \ldots, Y_p \in \mathcal{X}(M)$. We have*

$$
L_X(\varphi(Y_1, \ldots, Y_p)) = L_X(\varphi)(Y_1, \ldots, Y_p)
$$

$$
+ \sum_{i=1}^p \varphi(Y_1, \ldots, L_X Y_i, \ldots, Y_p).
$$

Proof: For clearness we write the calculation for $p = 2$ but the extension to any p is obvious. The first term of the righthandside gives :

$$\begin{aligned}
d \cdot L_X(\varphi)(Y,Z) &= (X_d)^*(\varphi)(Y,Z) - \varphi(Y,Z) \\
&= \varphi((X_{-d})^*Y, (X_{-d})^*Z) \circ X_d - \varphi(Y,Z) \\
&= \varphi((X_{-d})^*Y - Y, (X_{-d})^*Z) \circ X_d \\
&\quad + \varphi(Y, (X_{-d})^*Z - Z) \circ X_d \\
&\quad + \varphi(Y,Z) \circ X_d - \varphi(Y,Z) \\
&= \varphi(-dL_XY, (X_{-d})^*Z) \circ X_d \\
&\quad + \varphi(Y, -dL_XZ) \circ X_d + d \cdot L_X(\varphi(Y,Z)) \\
&= -\varphi(L_XY, d(X_{-d})^*Z) \circ X_d \\
&\quad - d \cdot \varphi(Y, L_XZ) \circ X_d + d \cdot L_X(\varphi(Y,Z)).
\end{aligned}$$

As $(X_{-d})^*Z = Z - d \cdot L_XZ$,

$$d \cdot (X_{-d})^*Z = d \cdot Z$$

and we get

$$\begin{aligned}
d \cdot L_X(\varphi)(Y,Z) &= -d \cdot \varphi(L_XY, Z) \circ X_d \\
&\quad - d \cdot \varphi(Y, L_XZ) \circ X_d + d \cdot L_X(\varphi(Y,Z)).
\end{aligned}$$

For any map f we have $(X_d)^*(f) = f + dL_X(f)$ and thus

$$d \cdot f \cdot X_d = d \cdot f$$

and so finally we get

$$\begin{aligned}
d \cdot L_X(\varphi)(Y,Z) &= -d \cdot \varphi(L_XY, Z) - d \cdot \varphi(Y, L_XZ) \\
&\quad + d \cdot L_X(\varphi(Y,Z))
\end{aligned}$$

for all d in D. Then

$$L_X(\varphi(Y,Z)) = L_X(\varphi)(Y,Z) + \varphi(L_XY, Z) + \varphi(Y, L_XZ)$$

as asked. ∎

d) In the same way, we can prove :

Proposition 5. *If* Φ *is in* $\mathcal{L}_R(\mathcal{X}(M)^p, \mathcal{X}(M))$,

$$L_X(\Phi(Y_1,\ldots,Y_p)) = L_X(\Phi)(Y_1,\ldots,Y_p) + \sum_i \Phi(Y_1,\ldots,L_XY_i,\ldots,Y_p).$$

■

In particular, this is an *a posteriori* justification of the definition of Lie derivative of p-forms as in 6.1.2, definition 7.

7.2.1   Lie groups and Lie algebras

In the description of the Lie algebra of vector fields (in section 3.2) we only used that $\mathcal{X}(M)$ is the tangent space at the identity, id_M, of the groups $\mathrm{Iso}(M)$ of bijective maps from M to M.

We make the simple but daring definition

Definition 2. *A Lie group is a microlinear group.*

In classical differential geometry, we must demand that multiplication and inversion are C^∞-mappings. In the spirit of synthetic differential geometry, all maps are viewed as differentiable. The definition thus becomes so simple. But there is here some act of carelessness. In fact, we thus consider the group $\mathrm{iso}(M)$ as a Lie group. But this is the correspondant in synthetic differential geometry of $\mathrm{Diff}^\infty(M)$ which classically is not a Lie group. The concept of Lie group is thus extended and in fact very much extended.

Let G be a Lie group. Let $\mathcal{G}$ the tangent space to G at the identity e of G. As any tangent space to a point of a microlinear object, $\mathcal{G}$ is a R-module. We define a bracket operation as follows. Let X and Y in $\mathcal{G}$. We consider

$$\tau : D \times D \to \mathcal{G}$$

given by

$$\tau(d_1, d_2) = X(d_1) \cdot Y(d_2) \cdot X(-d_1) \cdot Y(-d_2)$$

(the multiplication being in the group G). We have $\tau(d, o) = \tau(o, d) = e$ and thus, by the microlinearity of G, there is a unique factorization of τ through the multiplication $m : D \times D \to D$ (2.2.1 proposition 7).

The unique map from D to M so obtained is denoted by $[X, Y]$ and is characterized by

$$[X, Y](d_1 \cdot d_2) = \tau(d_1, d_2).$$

Just as in proposition 7 of 3.2.2 we obtain the

Proposition 6. *The described tangent space $\mathcal{G}$ with the described bracket is a R-Lie algebra.*

We say $\mathcal{G}$ is the Lie algebra of the group G.

It is possible to give a description of $\mathcal{G}$ as a subalgebra of $\mathcal{X}(G)$.

Definition 3. *Let Y be in $\mathcal{X}(G)$. We say Y is left-invariant if*

$$Y_d(h \cdot g) = h \cdot Y_d(g)$$

for all d in D, h and g in G.

Proposition 7. *The set of left-invariant vector fields on G is a sub-Lie-algebra of $\mathcal{X}(G)$ isomorphic with $\mathcal{G}$.*

Proof: First if $\underline{X}$ and $\underline{Y}$ are left-invariant vector fields such is also $[\underline{X}, \underline{Y}]$ as

$$
\begin{aligned}
[\underline{X}, \underline{Y}]_{d \cdot d'}(h \cdot g) &= \underline{Y}_{-d'}\left(\underline{X}_{-d}\left(\underline{Y}_{d'}(\underline{X}_d(h \cdot g))\right)\right) \\
&= h \cdot [\underline{X}, \underline{Y}]_{d \cdot d'}(g).
\end{aligned}
$$

Now, let $\underline{X}$ be a left-invariant vector field. We associate to it the tangent vector X at e given by $X(d) = \underline{X}_d(e)$. Conversely to $X : D \to G$ tangent at e we associate the vector field $\underline{X}$ defined by $\underline{X}_d(g) = g \cdot X(d)$. Trivially we have an isomorphism between the R-module of left-invariant vector fields and $\mathcal{G}$. Also it is an isomorphism of Lie algebras :

$$
\begin{aligned}
[\underline{X}, \underline{Y}]_{d \cdot d'}(g) &= \left(\underline{Y}_{-d'} \circ \underline{X}_{-d} \circ \underline{Y}_{d'} \circ \underline{X}_d\right)(g) \\
&= g \cdot \underline{Y}_{-d'}\left(\underline{X}_{-d}\left(\underline{Y}_{d'}(\underline{X}_d(e))\right)\right) \\
&= g \cdot \underline{X}_d(e) \cdot \underline{Y}_{-d'}\left(\underline{X}_{-d}(\underline{Y}_{d'}(e))\right)
\end{aligned}
$$

$$\begin{aligned}
&= \; g \cdot \underline{X}_d(e) \cdot \underline{Y}_{d'}(e) \cdot \underline{X}_{-d}(e) \cdot \underline{Y}_{-d'}(e) \\
&= \; g \cdot X(d) \cdot Y(d') \cdot X(-d) \cdot Y(-d') \\
&= \; g \cdot [X,Y](d \cdot d') \\
&= \; [X,Y]_{d \cdot d'}(g).
\end{aligned}$$

∎

7.2.2   Elementar linear examples

The basic example of a Lie group is a linear group. More precisely, let A be a Euclidean and microlinear R-algebra and E be a Euclidean and microlinear A-module. We write $GL_A(E)$ for the (microlinear) group of invertible elements of the R-module $\mathcal{L}_A(E)$. We have on $\mathcal{L}_A(E)$ the Lie algebra structure given by the commutator

$$[u,v] = u \circ v - v \circ u.$$

Proposition 8. *$\mathcal{L}_A(E)$ is the Lie algebra of the Lie group $GL_A(E)$.*

Proof: Let $X : D \to GL_A(E)$ be a tangent vector at id_E. We compose X with the inclusion of $GL_A(E)$ in $\mathcal{L}_A(E)$ which is Euclidean. Then we can write

$$X(d) = \; \mathrm{id}_E + d \cdot u$$

for a unique u in $\mathcal{L}_A(E)$. Conversely, given u we put $X(d) = \mathrm{id}_E + d \cdot u$. Clearly, $X(d)$ is invertible with $X(d)^{-1} = \mathrm{id}_E - d \cdot u = X(-d)$. These constructions are inverse and are R-linear. We show their compatibility with the Lie bracket. We consider X, Y tangent to $GL_A(E)$ at id_E with $X(d) = \mathrm{id}_E + d \cdot u$ and $Y(d) = \mathrm{id}_E + d \cdot v$.

$$\begin{aligned}
[X,Y](d_1 \cdot d_2) &= X(d_1) \cdot Y(d_2) \cdot X(-d_1) \cdot Y(-d_2) \\
&= (\mathrm{id}_E + d_1 \cdot u) \cdot (\mathrm{id}_E + d_2 \cdot v) \cdot (\mathrm{id}_E - d_1 \cdot u) \cdot (\mathrm{id}_E - d_2 \cdot v) \\
&= \mathrm{id}_E + d_1 d_2 (uv - vu) \\
&= \mathrm{id}_E + d_1 d_2 [u,v].
\end{aligned}$$

Then

$$[X,Y](d) = \; \mathrm{id}_E + d \cdot [u,v]$$

and the proof is complete.   ∎

Let A be a microlinear and Euclidean R-algebra and $\mathrm{Aut}(A)$ the microlinear group of automorphisms of A. Also we consider the set $\mathrm{Der}(A)$ of R-linear derivations, that is the set of R-linear mappings U from A to A such that :

$$U(a \cdot b) = U(a) \cdot b + a \cdot U(b).$$

Trivially the commutator $[U, V] = U \circ V - V \circ U$ of two derivations is a derivation. Then $\mathrm{Der}(A)$ is a sub-Lie-algebra of $\mathcal{L}_R(A)$.

Proposition 9. $\mathrm{Der}(A)$ *is the Lie algebra of the Lie group* $\mathrm{Aut}(A)$.

Proof: Let U in $\mathcal{L}_R(A)$. For all d in D we consider $\mathrm{id}_A + d \cdot U$. We have

$$\begin{aligned}
(\mathrm{id}_A + d \cdot U)(a) \cdot (\mathrm{id}_A + d \cdot U)(b) \\
= (a + d \cdot U(a)) \cdot (b + d \cdot U(b)) \\
= a \cdot b + d \cdot (a \cdot U(b) + U(a) \cdot b)
\end{aligned}$$

and

$$(\mathrm{id}_A + d \cdot U)(a \cdot b) = a \cdot b + d \cdot U(a \cdot b).$$

We see thus that U is a derivation if and only if, for all d in D, $\mathrm{id}_A + d \cdot U$ is an homomorphism from A to A and thus an automorphism of A, the inverse being $\mathrm{id}_A - d \cdot U$. ∎

We consider now an Euclidean, microlinear algebra A which is associative, commutative and unitary. Let E be a microlinear and Euclidean A-module. Let $h : E \times E \to A$ be a non-degenerate A-bilinear mapping. Let u be in $GL_A(E)$. We say that u is *h-orthogonal* if, for all X and Y in E,

$$h(u(X), u(Y)) = h(X, Y).$$

As classically, we can characterize the bijective h-orthogonal maps in another manner. Let u in $\mathcal{L}_A(E)$. We define the adjoint u^* of u by

$$h(X, u^*Y) = h(u(X), Y).$$

Of course u^* is well defined as h is non-degenerate, u^* is A-linear as u as h is A-bilinear. Then u is h-orthogonal if and only if

$$uu^* = u^*u = \mathrm{id}_E.$$

First if u is h-orthogonal and inversible,

$$h(X,Y) = h(u(X), u(Y)) = h\Big(X, u^*(u(Y))\Big)$$

and thus $u^*u = \mathrm{id}_E$. Then, as u is inversible, $u^* = u^{-1}$. Conversely

$$h(u(X), u(Y)) = h\Big(X, u^*(u(Y))\Big) = h(X,Y).$$

The set $\mathcal{O}_h(E)$ of h-orthogonal elements of $GL_A(E)$ is a Lie group.
We say that a A-linear map $U : E \to E$ is h-*antisymmetric* if for all X and Y in E

$$h(U(X), Y) + h(X, U(Y)) = 0.$$

It comes to the same to say $U + U^* = 0$. The set $\mathcal{L}_h(E)$ of h-antisymmetric, A-linear maps is a sub-Lie-algebra of $\mathcal{L}_A(E)$.

Proposition 10. $\mathcal{L}_h(E)$ *is the Lie algebra of* $\mathcal{O}_h(E)$.

Proof: Let U in $\mathcal{L}_A(E)$. For all d in D, we consider $\mathrm{id}_E + d \cdot U$. We have

$$h(X + d \cdot U(X), Y + d \cdot U(Y)) =$$
$$h(X,Y) + d \cdot (h(X, U(Y)) + h(U(X), Y)).$$

Then U is h-antisymmetric, i.e. in $\mathcal{L}_h(E)$ if and only if, for all d in D, $\mathrm{id}_E + d \cdot U$ is h-orthogonal. ∎

7.2.3 Global examples of Lie groups and Lie algebras

1) The first global example of the Lie algebra of a Lie group is $\mathcal{X}(M)$ itself which is the Lie algebra of the microlinear group $\mathrm{Iso}(M)$ of bijective maps from M to M.

2) Let
$$g : \mathcal{X}(M) \times \mathcal{X}(M) \to R^M$$
be a global Riemannian structure on M (7.1.1 definition 1). Now $\mathrm{Iso}(M,g)$ is the microlinear group of elements u in $\mathrm{Iso}(M)$ such that

$$u^*g = g.$$

We shall describe the Lie algebra of the Lie group $\mathrm{Iso}(M,g)$.

Definition 4. *A vector field X on M is said to be a* Killing vector field *on (M,g) if it has the derivation property :*

$$\forall Y, Z \in \mathcal{X}(M) \; L_X(g(Y,Z)) = g(L_XY, Z) + g(Y, L_XZ).$$

We denote by $K(M,g)$ the set of Killing vector fields on (M,g).

Proposition 11. *$K(M,g)$ is a sub-Lie-algebra of $\mathcal{X}(M)$ and it is the Lie algebra of the group $\mathrm{Iso}(M,g)$.*

Proof:

a) Let X_1 and X_2 be Killing vector fields. As X_1 and X_2 are Killing,

$$
\begin{aligned}
[X_1, X_2](g(Y,Z)) &= X_1\big(X_2(g(Y,Z))\big) - X_2\big(X_1(g(Y,Z))\big) \\
&= X_1(g([X_2,Y],Z)) + X_1(g(Y,[X_2,Z])) \\
&\quad - X_2(g([X_1,Y],Z)) - X_2(g(Y,[X_1,Z])) \\
&= g([X_1,[X_2,Y]],Z) + g([X_2,Y],[X_1,Z]) \\
&\quad + g([X_1,Y_1],[X_2,Z]) + g(Y,[X_1,[X_2,Z]]) \\
&\quad - g([X_2,[X_1,Y]],Z) - g([X_1,Y],[X_2,Z]) \\
&\quad - g([X_2,Y],[X_1,Z]) - g(Y,[X_2,[X_1,Z]]) \\
&= g([X_1,[X_2,Y]],Z) + g(Y,[X_1,[X_2,Z]]) \\
&\quad + g([X_2,[Y,X_1]],Z) + g(Y,[X_2,[Z,X_1]])
\end{aligned}
$$

and by the Jacobi identity, this becomes

$$[X_1, X_2](g(Y,Z)) = g([[X_1, X_2],Y],Z) + g(Y,[[X_1, X_2],Y])$$

then $[X_1, X_2]$ is a Killing vector field and thus $K(M,g)$ is a Lie algebra.

b) Let X in $\mathcal{X}(M)$. For all d in D, X_d is in $\mathrm{Iso}(M,g)$ if and only if for all d in D

$$(X_d)^*g = g$$

and then if and only if

$$L_X(g) = 0.$$

By proposition 4 (7.2.1) we have

$$L_X(g(Y,Z)) = L_X(g)(Y,Z) + g(L_XY,Z) + g(Y,L_XZ).$$

Then, $L_X(g) = 0$ if and only if X is a Killing vector field. So $K(M,g)$ is the Lie algebra of $\mathrm{Iso}(M,g)$.   ∎

3) The global pre-symplectic case (cfr. 7.1.2 definition 3) is similar to the Riemannian case. We can say that a vector field X is h-Killing (where h is the global pre-symplectic structure) if, for all Y and Z in $\mathcal{X}(M)$,
$$L_X(h(Y,Z)) = h(L_X(Y),Z) + h(Y,L_X(Z))$$
or, in other words,

$$X(h(Y,Z)) = h([X,Y],Z) + h(Y,[X,Z])$$

or, by proposition 4 in 7.2.1, if $L_X(h) = 0$. As in the Riemannian case, we say that $\mathrm{Iso}(M,h)$ is the microlinear group of elements u in $\mathrm{Iso}(M)$ such that
$$u^*h = h.$$
As in proposition 11, we can verify that the set $K(M,h)$ of h-Killing vector fields is a sub-Lie-algebra of $\mathcal{X}(M)$ and it is the Lie algebra of the group $\mathrm{Iso}(M,h)$.

In the symplectic case we have another description of the Lie algebra of $\mathrm{Iso}(M,h)$. We recall that if $h : \mathcal{X}(M) \times \mathcal{X}(M) \to R^M$ is a symplectic global structure on M, a vector field X is locally Hamiltonian if the 1-form i_X is closed (cfr. 7.1.2 definition 7).

Proposition 12. *The Lie algebra $LH(M,h)$ is the Lie algebra of the group $\mathrm{Iso}(M,h)$.*

Proof: Let X in $\mathcal{X}(M)$. For all d in D, $X_d \in \mathrm{Iso}(M,h)$ if and only if for all d in D, $(X_d)^*h = h$. Thus X takes his values in $\mathrm{Iso}(M,h)$ if and only if $L_Xh = 0$. But that is equivalent to

$$i_X(dh) + d(i_Xh) = 0.$$

As h is closed this gives $di_Xh = 0$ and thus X is locally Hamiltonian.∎

4) Let M be a microlinear object and ∇ a global connection on M.

Definition 5.

a) If $u \in \mathrm{Iso}(M)$, we say that u is an affine automorphism if

$$u^*(\nabla_X Y) = \nabla_{u^* X}(u^* Y)$$

for all X and Y in $\mathcal{X}(M)$. The set of these affine automorphism is denoted by $\mathrm{Iso}(M, \nabla)$.

b) If $X \in \mathcal{X}(M)$, we say that X is an affine vector field if

$$\nabla_{[X,Y]} = [L_X, \nabla_Y]$$

for all Y in $\mathcal{X}(M)$. The set of these affine vector field is denoted by $\mathcal{A}(M, \nabla)$.

Proposition 13. $\mathcal{A}(M, \nabla)$ is a sub-Lie-algebra of $\mathcal{X}(M)$ and it is the Lie algebra of the microlinear group $\mathrm{Iso}(M, \nabla)$.

Proof:

a) Let X and Y be in $\mathcal{A}(M, \nabla)$. We have

$$
\begin{aligned}
[L_{[X,Y]}, \nabla_Z] &= [[L_X, L_Y], \nabla_Z] \\
&= [L_X, [L_Y, \nabla_Z]] + [L_Y, [\nabla_Z, L_X]] \\
&= \nabla_{[X,[Y,Z]]} - \nabla_{[Y,[X,Z]]} \\
&= \nabla_{[[X,Y],Z]},
\end{aligned}
$$

and thus $[X, Y]$ is an affine vector field. Then $\mathcal{A}(M, \nabla)$ is a sub-Lie-algebra of $\mathcal{X}(M)$.

b) We consider $X : D \to \mathrm{Iso}(M, \nabla)$. For all d in D, X_d is an affine automorphism. We must show that X is affine.

$$
\begin{aligned}
d \cdot L_X(\nabla_Y Z) &= (X_d)^*(\nabla_Y Z) - \nabla_Y Z \\
&= \nabla_{(X_d)^* Y}((X_d)^* Z) - \nabla_Y Z \\
&= \nabla_{(X_d)^* Y}((X_d)^* Z - Z) + \nabla_{(X_d)^* Y} Z - \nabla_Y Z \\
&= \nabla_{(X_d)^* Y}(d L_X Z) + \nabla_{d \cdot L_X Y}(Z).
\end{aligned}
$$

As $dL_X Y = (X_d)^* Y - Y$, we have $d(X_d)^* Y = d \cdot Y$ and then

$$d \cdot L_X(\nabla_Y Z) = \nabla_{d \cdot Y}(L_X Z) + \nabla_{dL_X Y}(Z)$$

for all d in D. We have thus

$$L_X(\nabla_Y Z) = \nabla_Y(L_X Z) + \nabla_{[X,Y]}(Z)$$

and X is an affine vector field.

c) Conversely, let $X : D \to \mathrm{Iso}M$ be an affine vector field. We show that for all d in D, X_d is an affine automorphism. We have

$$(X_d)^*(\nabla_Y Z) = \nabla_Y Z + d \cdot L_X(\nabla_Y Z)$$

by the definition of L_X. In the same way

$$\nabla_{(X_d)^* Y}((X_d)^* Z) = \nabla_{Y + d \cdot L_X Y}(Z + d \cdot L_X Z)$$
$$= \nabla_Y Z + d(\nabla_Y(L_X Z) + \nabla_{[X,Y]} Z).$$

As X is affine we have

$$(X_d)^*(\nabla_Y Z) = \nabla_{(X_d)^* Y}((X_d)^* Z)$$

and thus X_d is an affine automorphism.   ∎

5) As a last example we consider the case of Poisson structures.

Definition 6.

a) Let (M_1, P_1) and (M_2, P_2) be two Poisson objects (cfr. 7.1.2 definition 9). A map $\varphi : M_1 \to M_2$ is a Poisson map if for all f and g in R^{M_2},
$$\{f \cdot g\}_2 \circ \varphi = \{f \circ \varphi, g \circ \varphi\}_1.$$

b) If (M, P) is a Poisson object, a vector field X is a Poisson vector field if it is in $\mathrm{Der}_p(A)$. In other words, X is a Poisson vector field if
$$X(\{f, g\}) = \{Xf, g\} + \{f, Xg\}.$$

We denote by Pois(M) the set of Poisson vector fields on M.

Consider Iso(M, P) $\subseteq$ Iso(M), the set of bijective maps which are Poisson map. It is a microlinear group. For example, if φ is a Poisson map, so is φ^{-1}. Because,

$$\{f, g\} \circ \varphi^{-1} = \{f \circ \varphi^{-1} \circ \varphi, g \circ \varphi^{-1} \circ \varphi\} \circ \varphi^{-1}$$
$$= \{f \circ \varphi^{-1}, g \circ \varphi^{-1}\} \circ \varphi \circ \varphi^{-1}.$$

Proposition 14. *Let* (M, P) *be a Poisson microlinear object. Then* Pois(M) *is the Lie algebra of* Iso(M, P).

Proof: Let $X : D \to$ Iso(M). For f, g in R^M and d in D,

$$\{f, g\} \circ X_d = \{f, g\} + d \cdot X(\{f, g\})$$

and

$$\{f \circ X_d, g \circ X_d\} = \{f + d \cdot X(f), g + d \cdot X(g)\}$$
$$= \{f, g\} + d(\{X(f), g\}, \{f, X(g)\}).$$

Then $X(\{f, g\}) = \{X(f), g\} + \{f, X(g)\}$, i.e. $X \in$ Der$_p A$ or $X \in$ Pois(M) if and only if for all d in D, X_d is in Iso(M, P). ∎

7.2.4   Actions of Lie groups

Let M be a microlinear object and G a Lie group with Lie algebra $\mathcal{G}$.

Definition 7.

1) *A left (resp. right) action of a Lie group* G *on* M *is an homomorphism (resp. antihomomorphism) from* G *to* Iso(M).

2) *A left (resp. right) action of* $\mathcal{G}$ *on* M *is an antihomomorphism (resp. homomorphism) from* $\mathcal{G}$ *to* $\mathcal{X}(M)$.

If α is a left action of G on M, $\alpha(g)(x)$ shall often be written $g * x$. For a right action β, we also write $x * g$. Of course, there is a bijection between left and right actions of G on M by $\beta(g) = \alpha(g^{-1})$. For the most part we only quote results for left actions.

Proposition 15. *If* $\alpha : G \to \mathrm{Iso}(M)$ *is a left action of G on M,*

$$T_e\alpha : \mathcal{G} = T_eG \to T_{\mathrm{id}}(\mathrm{Iso}M) = \mathcal{X}(M)$$

is a left action of $\mathcal{G}$ on M.

Proof: We compute $T_e\alpha([A,B])$ for A and B in $\mathcal{G}$. For $d_1, d_2 \in D$,

$$
\begin{aligned}
(T_e\alpha([A,B]))(d_1 \cdot d_2) &= \alpha([A,B](d_1 \cdot d_2)) \\
&= \alpha(A(d_1) \cdot B(d_2) \cdot A(-d_1) \cdot B(-d_2)) \\
&= T_e\alpha(A)(d_1) \circ T_e\alpha(B)(d_2) \circ T_e\alpha(A)(-d_1) \circ T_e\alpha(B)(-d_2) \\
&= -[T_e\alpha(A), T_e\alpha(B)](d_1 \cdot d_2).
\end{aligned}
$$

∎

Proposition 16. *A left action $\alpha : G \to \mathrm{Iso}(M)$ induces an homomorphism*

$$\bar{\alpha} : G \to \mathrm{Aut}(\mathcal{X}(M))$$

from G to the group of Lie algebra's automorphisms given by

$$\bar{\alpha}(g)(X)_d = \alpha(g) \circ X_d \circ \alpha(g^{-1}).$$

Proof: Trivially $\bar{\alpha}$ is a homomorphism from G to $\mathrm{Iso}(\mathcal{X}(M))$. But we must prove that $\bar{\alpha}(g) : \mathcal{X}(M) \to \mathcal{X}(M)$ is a homomorphism of Lie algebra. For linearity we prove homogeneity : for λ in R,

$$
\begin{aligned}
(\bar{\alpha}(g)(\lambda X))_d &= \alpha(g) \circ (\lambda X)_d \circ \alpha(g^{-1}) \\
&= \alpha(g) \circ X_{\lambda d} \circ \alpha(g^{-1}) \\
&= (\bar{\alpha}(g)(X))_{\lambda d} \\
&= (\lambda \cdot \bar{\alpha}(g)(X))_d.
\end{aligned}
$$

Next, for the bracket we have for $X, Y \in \mathcal{X}(M)$, $d_1, d_2 \in D$,

$$
\begin{aligned}
(\bar{\alpha}(g)([X,Y]))_{d_1 \cdot d_2} \\
&= \alpha(g) \circ Y_{-d_2} \circ X_{-d_1} \circ Y_{d_2} \circ X_{d_1} \circ \alpha(g^{-1}) \\
&= (\alpha(g) \circ Y_{-d_2} \circ \alpha(g^{-1})) \circ (\alpha(g) \circ X_{-d_1} \circ \alpha(g^{-1})) \\
&\quad \circ (\alpha(g) \circ Y_{d_2} \circ \alpha(g^{-1})) \circ (\alpha(g) \circ X_{d_1} \circ \alpha(g^{-1})) \\
&= [\bar{\alpha}(g)(X), \bar{\alpha}(g)(Y)]_{d_1 \cdot d_2}.
\end{aligned}
$$

∎

Proposition 17. *The map*

$$T_e\bar{\alpha} : \mathcal{G} = T_eG \longrightarrow T_{\mathrm{id}_{\mathcal{X}(M)}}(\mathrm{Aut}\mathcal{X}(M))$$

is an antihomomorphism of Lie algebras.

Proof: as for proposition 15. ∎

Note that, by proposition 9, $T_{\mathrm{id}_{\mathcal{X}(M)}}(\mathrm{Aut}\mathcal{X}(M))$ is isomorphic to $\mathrm{Der}(\mathcal{X}(M))$. We denote this isomorphism by λ :

$$T_{\mathrm{id}_{\mathcal{X}(M)}}(\mathrm{Aut}\mathcal{X}(M)) \xrightarrow{\lambda} \mathrm{Der}(\mathcal{X}(M)).$$

Definition 8. *If α is a left action of G on M, the adjoint ad_α of α is the antihomomorphism of Lie algebras*

$$\mathrm{ad}_\alpha = \lambda \circ T_e\bar{\alpha} : \mathcal{G} \to \mathrm{Der}(\mathcal{X}(M)).$$

Proposition 18. *For all A in $\mathcal{G}$ and X in $\mathcal{X}(M)$*

$$\mathrm{ad}_\alpha(A)(X) = [X, T_e\alpha(A)].$$

Proof: We have

$$
\begin{aligned}
\mathrm{ad}_\alpha(A)(X)_d &= \lambda(T_e\bar{\alpha}(A))(X)_d \\
&= \lambda(\bar{\alpha} \circ A)(X)_d
\end{aligned}
$$

and, by definition of λ (as in proposition 17)

$$(\bar{\alpha} \circ A)(d) = \mathrm{id}_{\mathcal{X}(M)} + d \cdot \lambda(\bar{\alpha} \circ A).$$

We have thus

$$
\begin{aligned}
\bar{\alpha}(A(d))(X)_{d_1} &= X_{d_1} \circ d \cdot \lambda(\bar{\alpha} \circ A)(X)_{d_1} \\
&= X_{d_1} \circ \lambda(\bar{\alpha} \circ A)(X)_{d \cdot d_1}.
\end{aligned}
$$

The definition of $\bar{\alpha}$ gives

$$\bar{\alpha}(A(d))(X)_{d_1} = \alpha(A(d)) \circ X_{d_1} \circ \alpha(A(-d))$$

and using these observations we get

$$
\begin{aligned}
\mathrm{ad}_\alpha(A)(X)_{d_1 \cdot d_2} &= \lambda(\bar{\alpha} \circ A)(X)_{d_1 \cdot d_2} \\
&= X_{d_1} \circ \bar{\alpha}(A(d))(X)_{d_1} \\
&= X_{d_1} \circ \alpha(A(d)) \circ X_{d_1} \circ \alpha(A(-d)) \\
&= [X, T_e\alpha(A)]_{d \cdot d_1}.
\end{aligned}
$$

∎

If, for example, we consider the identity action of the group $\text{Iso}(M)$ on M, the homomorphism $\overline{\text{id}} : \text{Iso}(M) \to \text{Aut}(\mathcal{X}(M))$ associates to φ the automorphism $\overline{\text{id}}(\varphi)$ described by

$$\overline{\text{id}}(\varphi)(X)_d = \varphi \circ X_d \circ \varphi^{-1}.$$

We thus have $\overline{\text{id}}(\varphi)(X) = \text{int}(\varphi) \circ X$ (where $\text{int}(\varphi)$ is the inner automorphism of $\text{Iso}(M)$ associated to φ). By the definition of ad_{id} :

$$\overline{\text{id}}(Y_d)(X) = X + d \cdot \text{ad}_{\text{id}}(Y)(X)$$

and, by proposition 18, the following follows.

Proposition 19. *For all X and Y in $\mathcal{X}(M)$ and d in D*

$$\text{int}(Y_d) \circ X = X + d \cdot [X, Y].$$

∎

Proposition 19 gives an infinitesimal relation between inner automorphisms and Lie brackets.

Now we describe the adjoint representation. Consider the action of G on G by inner automorphisms $\text{int} : G \to \text{Aut } G$. We denote by $\text{Ad}(g) : \mathcal{G} \to \mathcal{G}$ the map $T_e(\text{int}(g))$. We thus have

$$\text{Ad}(g)(A)_d = g \cdot A_d \cdot g^{-1}.$$

Proposition 20. *The map Ad is an homomorphism of G into the group $\text{Aut } \mathcal{G}$ of Lie automorphisms of $\mathcal{G}$.*

Proof: For example we show that $\text{Ad}(g)$ preserves Lie brackets :

$$\text{Ad}(g)([A, B])_{d_1 \cdot d_2} = g \cdot [A, B]_{d_1 \cdot d_2} \cdot g^{-1}$$
$$= g \cdot A_{d_1} \cdot g^{-1} \cdot g \cdot B_{d_2} \cdot g^{-1} \cdot g \cdot A_{-d_1} \cdot g^{-1} \cdot g \cdot B_{-d_2} \cdot g^{-1}$$
$$= [\text{Ad}(g)(A), \text{Ad}(g)(B)]_{d_1 \cdot d_2}.$$

∎

There is another description of Ad in terms of right translations. Let $\alpha : G \to$ Iso G defined by $\alpha(g)(h) = hg^{-1}$. By proposition 16, $\bar{\alpha} : G \to \mathrm{Aut}(\mathcal{X}(G))$ is given by

$$
\begin{aligned}
\bar{\alpha}(g)(X)_d(h) &= (\alpha(g) \circ X_d \circ \alpha(g^{-1}))(h) \\
&= X_d(hg) \cdot g^{-1}.
\end{aligned}
$$

If X is left invariant, so is $\bar{\alpha}(g)(X)$. We get an homomorphism, also written $\bar{\alpha}$, from G to Aut $\mathcal{G}$.

Proposition 21. *With above notations $\bar{\alpha} = \mathrm{Ad}$.*

Proof:

$$
\begin{aligned}
\bar{\alpha}(g)(A)_d &= \bar{\alpha}(g)(\bar{A})_d(e) \\
&= \bar{A}_d(g) \cdot g^{-1} \\
&= g \cdot A_d \cdot g^{-1}.
\end{aligned}
$$

∎

Definition 9. *The* adjoint representation *of $\mathcal{G}$ into* Der $\mathcal{G}$ *is the composition*

$$
\mathrm{ad} = \lambda \circ T_e \mathrm{Ad}.
$$

Proposition 22. *The adjoint representation* ad *is the Lie inner representation in that sense that*

$$
\mathrm{ad}(A)(B) = [A, B].
$$

Proof:

$$
\begin{aligned}
\mathrm{ad}(A)(B) &= \lambda(T_e \mathrm{Ad}(A))(B) \\
&= \lambda(T_e \bar{\alpha}(A))(B) \\
&= [B, T_e \alpha(A)]
\end{aligned}
$$

by proposition 18. We have

$$
(T_e \alpha)(A)_d(g) = \alpha(A_d)(g) = g \cdot A_{-d},
$$

then

$$
(T_e \alpha)(A) = -A
$$

and thus

$$
\mathrm{ad}(A)(B) = [B, -A] = [A, B].
$$

∎

We finish with the concept of *coadjoint representation*. Let E be a R-module and $\zeta : G \to GL(E)$ an homomorphism named a linear representation of G in E. It corresponds to it a linear representation, ζ^*, of G in $E^* = \mathcal{L}(E, R)$ named the contragredient of ζ and described by

$$\zeta^*(g)(\alpha)(v) = \alpha(\zeta(g^{-1})(v))$$

for all α in E^* and v in E.

The contragredient Ad^* of Ad is the coadjoint representation of G. The coadjoint representation of $\mathcal{G}$ is given by

$$ad^* = \lambda \circ T_e(Ad^*)$$

(where λ is the isomorphism of $T_{id}(GL(\mathcal{G}^*))$ onto the Lie algebra $\mathcal{L}(\mathcal{G}^*, \mathcal{G}^*)$. For convenience we write $\langle \alpha, v \rangle$ for $\alpha(v)$.

Proposition 23. *The coadjoint representation* ad^* *of* $\mathcal{G}$ *is characterized by*

$$\langle ad^*(A)(\alpha), B \rangle = -\langle \alpha, [A, B] \rangle$$

for A and B in $\mathcal{G}$ and α in $\mathcal{G}^*$.

   Proof: As

$$Ad^*(A_d) = id_{\mathcal{G}^*} + d \cdot ad^*(A),$$

for all d in D

$$\begin{aligned}
\langle \alpha + d \cdot ad^*(A)(\alpha), B \rangle &= \langle A^*(A_d)(\alpha), B \rangle \\
&= \langle \alpha, Ad(A_{-d})(B) \rangle \\
&= \langle \alpha, B - d \cdot ad(A)(B) \rangle \\
&= \langle \alpha, B \rangle - d \cdot \langle \alpha, [A, B] \rangle.
\end{aligned}$$

Then, the coefficients of d are equal and

$$\langle ad^*(A)(\alpha), B \rangle = -\langle \alpha, [A, B] \rangle.$$

■

7.3   The cotangent bundle

7.3.1   Vector fields on a vector bundle

Let M be a microlinear object and $\pi : E \to M$ a Euclidean and microlinear vector bundle over M.

Definition 1. *A vector field based on π is a pair $(\bar{X}, X)$, such that $\bar{X} \in \mathcal{X}(E)$, $X \in \mathcal{X}(M)$ and the following square commutes*

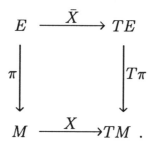

We denote by $\mathcal{X}_\pi(E)$ the set of vector fields based on π.

As π is surjective, X is unique in that sense that $\iota : \mathcal{X}_\pi(E) \to \mathcal{X}(E) : (\bar{X}, X) \to \bar{X}$ is injective. Thus $\mathcal{X}_\pi(E)$ can be seen as a Lie sub-algebra of $\mathcal{X}(E)$.

We denote by $\mathrm{Iso}_\pi(E)$ the set of pairs (φ, f) in $\mathrm{Iso}(E) \times \mathrm{Iso}(M)$ such that $\pi \circ \varphi = f \circ \pi$. The function $j : \mathrm{Iso}_\pi(E) \to \mathrm{Iso}(E) : (\varphi, f) \to \varphi$ is injective. Thus $\mathrm{Iso}_\pi(E)$ can be seen as a subgroup of $\mathrm{Iso}(E)$. We trivially have the following proposition.

Proposition 1. $\mathcal{X}_\pi(E)$ *is the Lie algebra of* $\mathrm{Iso}_\pi(E)$.   ∎

We also consider the set $\mathrm{Iso}_0(E)$ of (φ, id_M) in $\mathrm{Iso}_\pi(E)$ and the set $\mathcal{X}_0(E)$ of vertical vector fields, that is the $(\bar{X}, \mathcal{O}_M)$ with null projection in $\mathcal{X}(M)$.

Proposition 2. $\mathrm{Iso}_0(E)$ *is a normal subgroup of* $\mathrm{Iso}_\pi(E)$ *;* $\mathcal{X}_0(E)$ *is an ideal of the Lie algebra* $\mathcal{X}_\pi(E)$ *and* $\mathcal{X}_0(E)$ *is the Lie algebra of* $\mathrm{Iso}_0 E$.

Proof: For example, we have

$$(\psi, g) \circ (\varphi, \mathrm{id}_M) \circ (\psi^{-1}, g^{-1}) = (\psi \circ \varphi \circ \psi^{-1}, \mathrm{id}_M)$$

and

$$[(\bar{X},0),(\bar{Y},Y)] = ([\bar{X},\bar{Y}],[0,Y]) = ([\bar{X},\bar{Y}],0).$$

∎

There is another characterization of vertical vector fields. Let $\mathcal{F}(E)$ be the set of maps $S : E \to E$ such that $\pi \circ S = \pi$.

Proposition 3. *There is a bijective map from $\mathcal{F}(E)$ onto $\mathcal{X}_0(E)$ sending $S \in \mathcal{F}(E)$ to the vector field V_S given by*

$$V_S(\alpha)(d) = \alpha + d \cdot S(\alpha)$$

for all α in E and d in D.

Proof: If S is in $\mathcal{F}(E)$, obviously V_S is a vertical field. Conversely, if V is a vertical field then for each α in E, $V(\alpha)$ is tangent in α to the fibre of E over $x = \pi(\alpha)$. As this fibre E_x is Euclidean, there is a unique element $S_V(\alpha)$ in E_x such that

$$V(\alpha)(d) = \alpha + d \cdot S_V(\alpha).$$

∎

Definition 2. *The Liouville's field on $\pi : E \to M$ is the vertical field Z corresponding to the identity on E. Explicitely*

$$Z(\alpha)(d) = \alpha + d \cdot \alpha.$$

Also we can extend the concept of connection (chapter 5) into a concept of connection over the vector bundle $\pi : E \to M$.

Definition 3. *A connection on a vector bundle $\pi : E \to M$ is a map*

$$\nabla : E \underset{M}{\times} TM \to M$$

such that

1) $\nabla(\alpha,t)(o) = \alpha$
    $\pi \circ \nabla(\alpha,t) = t$
2) $\nabla(\lambda \cdot \alpha, t)(d) = \lambda \cdot \nabla(\alpha,t)(d)$
    $\nabla(\alpha, \lambda t)(d) = \nabla(\alpha,t)(\lambda d)$

for all $\alpha \in E$, $t \in TM$ (with $\pi(\alpha) = t(o)$), $\lambda \in R$ and $d \in D$.

We consider the map

$$K : TE \to E \underset{M}{\times} TM : v \mapsto (v(o), \pi \circ v).$$

The condition (1) of definition 3 says that ∇ is a section of K. Condition (2) express bilinearity for R-module structures on the fibers of E and those of TM. We also consider the map

$$\mathcal{V} : E \underset{M}{\times} E \to TE : (\alpha, \beta) \mapsto \alpha + d\beta.$$

Proposition 4. *We take a connection ∇ on a Euclidean vector bundle $\pi : E \to M$. The sequence*

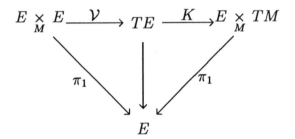

is a short exact sequence of vector bundles over E.

Proof:

a) We know that $\mathcal{V}$ is injective and that for each α in E, the map $\mathcal{V}(\alpha, \beta) = \alpha + d\beta$ is linear from E into $T_\alpha E$.

b) K is linear on fibre as, for each α, K send $v \in T_\alpha E$ on $(\alpha, \pi \circ v)$. On the other side by hypothesis K has a section ∇.

c) Lastly an element v in $T_\alpha E$ is vertical if $K(v) = (\alpha, o_x)$ ($x = \pi(\alpha)$). But that occurs if and only if there exists β in E such that $v = \alpha + d\beta$.

■

Thus we are in a situation generalizing that of section 5.2. For example, for each α, $T_\alpha E$ is the direct sum of

$$V_\alpha E = \{\mathcal{V}(\alpha, \beta)|\beta \in E\}$$

and

$$H_\alpha E = \{\nabla(\alpha, t)|t \in T_x M\}.$$

As in 5.2.2, we can describe the connecting map from TE into E. If v is in E, $v - \nabla K v$ is a vertical vector. There is a unique element $C(v)$ in E which is the derivation of $v - \nabla K v$.

Definition 4. *The connecting map $C : TE \to E$ associated to the connection ∇ on $\pi : E \to M$ is the map such that for all v in E*

$$v - \nabla K v = v(o) + \underset{\sim}{d} \cdot C(v).$$

The following trivial observation is usefull.

Proposition 5. *Let ∇ be a connection on $\pi : E \to M$. Every vector field on M has a unique horizontal lift for ∇.*

Proof: If X is in $\mathcal{X}(M)$, we define $\bar{X} \in \mathcal{X}(E)$ by

$$\bar{X}_\alpha = \nabla(\alpha, X_\alpha).$$

■

We also can examine things from a global point of view. We consider the A-module $\Gamma(E)$ of sections of $\pi : E \to M$. Then we have a notion of global connection in term of covariant derivation.

Definition 5. *A covariant derivation on a vector bundle $\pi : E \to M$ is a connection on the A-module $\Gamma(E)$ (cf. section 6.1.3, definition 9). More explicitely, it is a R-bilinear map*

$$\nabla : \mathcal{X}(M) \times \Gamma(E) \to \Gamma(E) : (X, A) \mapsto \nabla_X A$$

such that

(1) $\nabla_{a \cdot X} A = a \cdot \nabla_X A$
(2) $\nabla_X (a \cdot A) = a \cdot \nabla_X A + X(a) \cdot A.$

Let $\nabla : E \underset{M}{\times} TM \to TM$ be a connection on π (definition 3), X in $\mathcal{X}(M)$ and A in $\Gamma(E)$. For any x in M, we put $\alpha = A(x)$. We consider in $T_\alpha E$ the difference

$$A(X(x,\underline{d})) - \nabla(A(x), X(x,\underline{d})).$$

His projection is zero, then it is vertical. We can express the following proposition.

Proposition 6. *If ∇ is a connection on $\pi : E \to M$, the map from $\mathcal{X}(M) \times \Gamma(E)$ into $\Gamma(E)$ which associates to (X, A), the section $\nabla_X A$ characterized by*

$$A(X(x,\underline{d})) - \nabla(A(x), X(x,\underline{d})) = A(x) + \underline{d}\nabla_X A(x)$$

is a covariant derivation on π.

Proof: Immediately, we have $\nabla_{a \cdot X} = a \cdot \nabla_X A$. On the other side

$$
\begin{aligned}
(a \cdot A)(X(x,\underline{d})) &- \nabla((a \cdot A)(x), X(x,\underline{d})) \\
&= a(X(x,\underline{d})) \cdot A(X(x,\underline{d})) - a(x) \cdot \nabla(A(x), X(x,\underline{d})).
\end{aligned}
$$

As $a(X(x,d)) = a(x) + d \cdot X(a)(x)$, we obtain

$$
\begin{aligned}
((a \cdot A)(X_x) &- \nabla((a \cdot A)(x), X_x))(d) \\
&= a(x) \cdot (A(X_x) - \nabla(A(x), X_x))(d) \\
&\quad + d \cdot X(a)(x) \cdot A(X(x,d)) \\
&= (A(X_x) - \nabla(A(x), X_x))(a(x) \cdot d) + d \cdot X(a)(x)A(x) \\
&= A(x) + a(x) \cdot d \cdot \nabla_X A(x) + d \cdot X(a)(x)A(x).
\end{aligned}
$$

Thus, we have

$$\nabla_X(a \cdot A) = a \cdot \nabla_X A + X(a) \cdot A.$$

∎

7.3.2   Symplectic structure on cotangent bundle

Since the work of Poincaré, it can be said that a dynamical system is a vector field on a phase space which is a differentiable manifold. But the present description of Hamiltonian mechanics becomes that of a Hamiltonian vector field on a symplectic manifold. This view comes from classical dynamic when it has been observe that Liouville 2-form gives a symplectic structure on the cotangent bundle.

Of course, nobody can hope to find here a synthetic version of a actually substantial part of mechanics. I think it is nevertheless interesting to describe the Liouville forms and the symplectic structure on the cotangent bundle.

Let M be a microlinear object and g a pointwise Riemannian structure on M. We can consider the cotangent bundle $\pi : T^*M \rightarrow M$ whose fibres are $T_x^*M = (T_xM)^*$. These fibres are isomorphic to T_xM thanks to g. Thus T^*M is a microlinear and Euclidean vector bundle.

As for any vector bundle, TT^*M has two vector bundle structures. First,

$$TT^*M \rightarrow T^*M : v \mapsto v(o)$$

is the usual structure of tangent bundle. Secondly

$$TT^*M \rightarrow TM : v \mapsto \pi \circ v$$

arises from the fact that $\pi : T^*M \rightarrow M$ is itself a vector bundle.

What is peculiar to the cotangent bundle, are the Liouville forms.

Definition 6. *The Liouville* 1-*form on* $\pi : T^*M \rightarrow M$ *is the* 1-*form*

$$\theta : TT^*M \rightarrow R$$

given by

$$\theta(v) = v(o)(\pi \circ v)$$

for any v *in* TT^*M. *The Liouville* 2-*form on* π *is* $\omega = -d\theta$.

It is interesting to note that θ has a double linearity.

Proposition 7. *The Liouville* 1-*form* $\theta : TT^*M \rightarrow R$ *is linear for each of two vector bundle structures on* TT^*M.

Proof: We denote by λv the product $(\lambda v)(d) = v(\lambda d)$ - that is the product on the fibre of $TT^*M \to T^*M$. We have

$$
\begin{aligned}
\theta(\lambda v) &= (\lambda v)(o)(\pi \circ \lambda v) \\
&= v(o)(\lambda(\pi \circ v)) \\
&= \lambda\theta(v).
\end{aligned}
$$

We denote by $\lambda \cdot v$ the product $(\lambda \cdot v)(d) = \lambda \cdot v(d)$ - that is the product in the fibre of $T\pi : TT^*M \to TM$. We have

$$
\begin{aligned}
\theta(\lambda \cdot v) &= (\lambda \cdot v)(o)(\pi \circ (\lambda \cdot v)) \\
&= \lambda v(o)(\pi \circ v) \\
&= \lambda\theta(v).
\end{aligned}
$$

∎

As usual the pointwise forms θ and ω induce global forms on the $\mathcal{X}(T^*M)$ but also on the $\mathcal{X}_\pi(T^*M)$.

If we have a connection on M

$$
\nabla : TM \underset{M}{\times} TM \to TTM
$$

in the sense of chapter 5, we can use the isomorphism $\underline{g} : TM \to T^*M$ to build a connection ∇^* on T^*M making the diagram

$$
\begin{array}{ccc}
TM \underset{M}{\times} TM & \xrightarrow{\nabla} & TTM \\
{\scriptstyle \underline{g} \times 1}\downarrow & & \downarrow{\scriptstyle T\bar{g}} \\
T^*M \underset{M}{\times} TM & \xrightarrow{\nabla^*} & TT^*M
\end{array}
$$

to commute. Explicitly,

$$
\nabla^*(\alpha, t)(d) = \underline{g}(\nabla(\underline{g}^{-1}(\alpha), t)(d', d))
$$

for $t \in T_x M$, $\alpha \in T_x^* M$, $d \in D$.

We also can look things from a global point of view. If we have a global covariant derivation

$$\mathcal{X}(M) \times \mathcal{X}(M) \to \mathcal{X}(M) : (X, Y) \mapsto \nabla_X Y,$$

composition with $\underline{g} : TM \to T^*M$ transform elements of $\mathcal{X}(M)$ into sections in $\Gamma(T^*M)$. We obtain a covariant derivation ∇^* on π such that the following square commutes

$$
\begin{array}{ccc}
\mathcal{X}(M) \times \mathcal{X}(M) & \xrightarrow{\nabla} & \mathcal{X}(M) \\
{\scriptstyle 1 \times (\underline{g} \circ -)} \downarrow & & \downarrow {\scriptstyle \underline{g} \circ -} \\
\mathcal{X}(M) \times \Gamma(T^*M) & \xrightarrow{\nabla^*} & \Gamma(T^*M)
\end{array}
$$

We have now the basic result.

Proposition 8. *Let (M, g) be a microlinear object with a pointwise Riemannian structure and let ∇ be a connection on M. The Liouville 2-form ω is a symplectic structure on $\mathcal{X}_\pi(T^*M)$.*

Proof: It is clear that ω is a closed 2-form because it is exact ; $\omega = -d\theta$. We must show that ω is non degenerate. Let V be in $\mathcal{X}_\pi(T^*M)$ and we suppose that for any W in $\mathcal{X}_\pi(T^*M)$, $\omega(V, W) = 0$.

Thus we have

$$0 = -\omega(V, W) = d\theta(V, W) = V(\theta(W)) - W(\theta(V)) - \theta([V, W]).$$

If W is vertical, $[V, W]$ also is vertical (cf. proposition 2). If W is vertical, $\pi \circ W = 0$ and then $\theta(W) = 0$. We get

$$0 = \omega(V, W) = W(\theta(V)).$$

For α in T^*M we have

$$0 = d \cdot W(\theta(V))(\alpha) = \theta(V)(W(\alpha, d)) - \theta(V)(\alpha).$$

Now we take $X \in \mathcal{X}(M)$ and we have $X^* = \underline{g} \circ X : M \to T^*M$. We consider the vertical vector field associated to that section X^*, i.e.

$$W(\alpha, d) = \alpha + d \cdot X^*(\pi(\alpha)).$$

Thus,

$$\begin{aligned}
0 &= \theta(V)(W(\alpha, d)) - \theta(V)(\alpha) \\
&= \theta(V)\Big(\alpha + d \cdot X^*(\pi(\alpha))\Big) - \theta(V)(\alpha) \\
&= \theta(V_{\alpha + d \cdot X^*(\pi(\alpha))}) - \theta(V_\alpha) \\
&= V_{\alpha + d \cdot X^*_{\pi(\alpha)}}(o)(\pi \circ V_{\alpha + d \cdot X^*_{\pi(\alpha)}}) - \theta(V_\alpha) \\
&= \Big(\alpha + d \cdot X^*(\pi(\alpha))\Big)(\pi \circ V_{\alpha + d \cdot X^*(\pi(\alpha))}) - \theta(V_\alpha).
\end{aligned}$$

But V is based on π and thus $\pi \circ V_{\alpha + d \cdot X^*(\pi(\alpha))} = \pi \circ V_\alpha$ and so

$$\begin{aligned}
0 &= \alpha(\pi \circ V_\alpha) + d \cdot X^*_{\pi(\alpha)}(\pi \circ V_\alpha) - \theta(V_\alpha) \\
&= d \cdot X^*_{\pi(\alpha)}(\pi \circ V_\alpha)
\end{aligned}$$

for all d in D. Then

$$X^*_{\pi(\alpha)}(\pi \circ V_\alpha) = 0$$

i.e., by the definition of X^*,

$$g(X, \pi \circ V) = 0$$

for all X in $\mathcal{X}(M)$. As g is nondegenerate

$$\pi \circ V = 0$$

and thus V is a vertical vector field.

We write this vertical field V as

$$V_\alpha(d) = \alpha + d \cdot C(V)(\alpha).$$

Then we take a horizontal W. More precisely, let Z in $\mathcal{X}(M)$ and $Z^* = \underline{g} \circ Z$. We take

$$W_\alpha = \nabla(Z^*_x, Z_x).$$

We have

$$
\begin{aligned}
0 &= -d \cdot \omega(V, W)(\alpha) \\
&= \theta(W)(V_\alpha(d)) - \theta(W)(\alpha) \\
&= d \cdot \theta(V)(C(V)(\alpha)) \\
&= d \cdot C(V)(\alpha)(\pi \circ W).
\end{aligned}
$$

But $\pi \circ W = \pi \circ \nabla(Z^*, Z) = Z$. Then $C(V)(\alpha)(Z) = 0$ for all Z and all α and thus $C(V) = 0$.

As we have $\pi \circ V = 0$ and $C(V) = 0$, V is zero and $\omega = -d\theta$ is non degenerate. ∎

The Liouville vector field on T^*M (cf. definition 2) can be characterized as follows.

Proposition 9. *The Liouville's field Z on $\pi : T^*M \to M$ is the unique vector field such that*

$$
i_Z(\omega) = -\theta.
$$

Proof: We consider the Liouville's field

$$
Z_\alpha = \alpha + d\alpha.
$$

For all $V \in \mathcal{X}_\pi(T^*M)$

$$
\begin{aligned}
i_Z(\omega)(V) &= -d\theta(Z, V) \\
&= -Z(\theta(V))
\end{aligned}
$$

because Z and also $[Z, V]$ are vertical. We get

$$
\begin{aligned}
d \cdot Z(\theta(V))(\alpha) &= \theta(V)(\alpha + d \cdot \alpha) - \theta(V)(\alpha) \\
&= \theta(V_{\alpha + d \cdot \alpha}) - \theta(V_\alpha) \\
&= V_{\alpha + d \cdot \alpha}(o)(\pi \circ V_{\alpha + d \cdot \alpha}) - \theta(V_\alpha) \\
&= (\alpha + d \cdot \alpha)(\pi \circ V_\alpha) - \theta(V_\alpha) \\
&= d \cdot \alpha(\pi \circ V_\alpha) \\
&= d \cdot \theta(V)(\alpha)
\end{aligned}
$$

and $i_Z(\omega)(V) = -Z(\theta(V)) = -\theta(V)$.

Conversely, as $\omega = -d\theta$ is non degenerate, $\omega(Z, V) = -\theta(V)$ characterizes Z. ∎

It would be interesting to have a pointwise symplectic structure on the cotangent bundle. For that we need another structure on M.

Definition 7. *Let M be a microlinear object. A diffusion on M consists, for each x in M, in a linear map $P_x : T_x M \to \mathcal{X}(M)$ such that $P_x(t)(x) = t$.*

If $\pi : E \to M$ is a vector bundle (microlinear and Euclidean), a *diffusion* on π consists, for each α in E, in a linear map $P_\alpha : T_\alpha E \to \mathcal{X}_\pi(E)$ such that $P_\alpha(v)(\alpha) = v$.

If $M = R^n$ we have a canonical diffusion on R^n. In fact, for $x \in R^n$ and $t = x + \underline{d}y$ in $T_x R^n$ we simply put

$$P_x(x + \underline{d}y)(u) = u + \underline{d}y$$

for all u in R^n. In other words, we associate to $x + \underline{d}y$ the parallel constant field on R^n.

In classical differential geometry paracompact manifolds have diffusions. We can consider, for a tangent vector at x in M, the lift of a constant vector field in a local chart and extend this using a flat function with support in the domain of the chart. This construction uses choice and is not intuitionnistic.

Proposition 10. *Let $\pi : E \to M$ be a Euclidean vector bundle with connection ∇. If P is a diffusion on M, we can construct a diffusion $\bar{P}$ on π.*

Proof: We take $\alpha \in E$, $v \in T_\alpha E$ with $\pi(\alpha) = x$. We have a vector field on M

$$X = P_x(\pi \circ v).$$

The vertical component of v is $\alpha + \underline{d}C(v)$. We have the vertical field Y on E given by $Y(\beta) = \beta + \underline{d}C(v)$. And the diffusion $\bar{P}$ on π is given by

$$\bar{P}_\alpha(v) = X + Y.$$

∎

Proposition 11. *Let (M, g) be a microlinear object with a pointwise Riemannian structure, ∇ a connection on M and P a diffusion on M. The map*

$$h : TT^*M \underset{T^*M}{\times} TT^*M \to R$$

defined by

$$h_\alpha(v, w) = \omega(\bar{P}_\alpha v, \bar{P}_\alpha w)(\alpha)$$

where $\bar{P}$ is the diffusion on π associated to P is a pointwise symplectic structure on T^*M.

Proof: Clearly h_α is bilinear and antisymmetric. To prove that it is non degenerate we can mimic the proof of proposition 8 using P and $\bar{P}$. ∎

7.3.3   The case of R^n

Breefly we consider the case $M = R^n$, just to write down the classical Hamilton equations. We identify TR^n with $R^n \times R^n$ by the bijective map

$$R^n \times R^n \to TR^n : (x, u) \mapsto x + \underline{d}u.$$

We denote by q_i $(1 \le i \le n)$ the first n coordinates and by p_i $(1 \le i \le n)$ the last ones. Of course we have on R^n the canonical scalar product and also the canonical diffusion. We have also the flat connection on R^n.

Proposition 12. *The Liouville's 1-form* $\theta : TT^*R^n \to R$ *is*

$$\theta = \sum_{i=1}^{n} p_i dq_i$$

and thus the Liouville's 2-form is

$$\omega = -\sum_{i=1}^{n} dp_i \wedge dq_i = \sum_{i=1}^{n} dq_i \wedge dp_i.$$

Proof: Let (x, u) a point in T^*R^n and

$$\underline{w} = (x, u) + \underline{d}(y, v)$$

a tangent vector to T^*R^n at (x, u). We have $\pi \circ \underline{w} = x + \underline{d}y$ and $w(o) = (x, u)$. The definition of θ (7.2.2 definition 6) gives

$$\theta(\underline{w}) \;=\; \underline{w}(o)(\pi \circ \underline{w})$$

$$\begin{aligned} &= (x, u)(x + \underline{d}y) \\ &= u(y) \\ &= \sum_{i=1}^{n} u_i y_i \end{aligned}$$

(we identify u with the linear form du). On another side we have

$$\left(\sum_{i=1}^{n} p_i dq_i \right)_{(x,u)} (x + \underline{d}y)$$

$$= \sum_{i=1}^{n} u_i \cdot dq_i(x + \underline{d}y)$$

$$= \sum_{i=1}^{n} u_i y_i$$

and then $\theta = \Sigma p_i dq_i$. Thus also $\omega = -\Sigma dp_i \wedge dq_i = \Sigma dq_i \wedge dp_i$.   ∎

In the present situation we return to Hamiltonian field (for the general definition see 7.1.2 definition 7). Let $H : T^*R^n \to R$ an arbitrary map which we can name a Hamiltonian. His differential dH is a 1-form on T^*R^n. As ω is non degenerate, there exists a unique vector field X_H, Hamiltonian for H, such that for all Y

$$dH(Y) = \omega(X_H, Y).$$

In the coordinates (q_i, p_i) we have

$$dH = \sum_{i=1}^{n} \frac{\partial H}{\partial q_i} dq_i + \sum_{i=1}^{n} \frac{\partial H}{\partial p_i} dp_i.$$

We look for coordinates (A_i, B_i) of the field X_H. We have

$$\begin{aligned} dH &= i_{X_H}(\omega) = i_{X_H}(dq_i \wedge dp_i) \\ &= dq_i(X_H) dp_i - dp_i(X_H) dq_i. \end{aligned}$$

But $dq_i(X_H) = A_i$ and $dp_i(X_H) = B_i$ and thus

$$A_i = \frac{\partial H}{\partial p_i}, \quad B_i = -\frac{\partial H}{\partial q_i}.$$

Then we have :

Proposition 13. *A curve $(q_i(t), p_i(t))$ in T^*R^n is an integral curve for the Hamiltonian field X_H if and only if it verifies the Hamilton equations*

$$\frac{dq_i}{dt} = \frac{\partial H}{\partial p_i} \quad and \quad \frac{dp_i}{dt} = -\frac{\partial H}{\partial q_i}.$$

∎

7.4 Commented bibliography

Here the good references are to classical books on differential geometry applied to mechanics. For example we must recall the books of Abraham and Marsden [1], Godbillon [21], and Liberman ans Marle [45]. The concept of structured Lie object is only an algebrization of classical situations, adapted to S.D.G.

The present synthesis on general Lie groups appears in Lavendhomme [43](1994). Other topics are new in the synthetic context but old in the classical context.

Other aspects of differential geometry are also developed in the synthetic way, for example in Bunge [6] (1983), Bunge and Dubuc [8] (1987), Bunge and Gago [9] (1988), or Gago [19] (1988) and [20] (1989).

Chapter 8

Note on toposes and models of S.D.G.

Here our aim is very different to that of the preceding chapters. In a sense our account of S.D.G. is finished. Now our objective is to show how it is possible to understand the idea of an intuitionistic set theory, or, more precisely, the concept of toposes. We shall also indicate some models of S.D.G., so showing how results can be applied to classical differential geometry.

In the preceding chapters we gave complete proofs ; Here we shall not, but all missing proofs can be found in two works. For categories and toposes we refer to the handbook of categorical algebra of F. Borceux [5] and for models of S.D.G. to the book of I. Moerdijk and G. E. Reyes [52].

8.1 Multisorted language of higher order

8.1.1 The language

We shall use typed languages : for each variable in the language will be associated a type. Such "typings" appear often in mathematics. For example when in linear algebra, we write

$$(\alpha + \beta)x = \alpha x + \beta x$$

the α and β are of type "scalar" and x of type "vector" . With simple types we can construct more complex ones. For example in the expression $x \in X \in \mathcal{X}$ it is understood that x is of a given type, say of type "point" , but X is of type "set of points" and $\mathcal{X}$ of type "set of set of points". Similarly operations and relations have relations with types. For example the expression αx refers to an operation with ("scalar", "vector") as source type and "vector" as target type.

A more formal definition is the following.

A multisorted language L of higher order is given by the following data :

(a) A class Typ called the class of *types*. We assume certain structure on Typ :

   (1) A constant type writen 1 ;

   (2) A binary law which associates to types X and Y a type (X, Y) ;

   (3) A unary law which to a type X associates a type denoted by $P(X)$. We write Ω for $P(1)$

(b) For each type X, we give us an infinite set V_X whose elements are the *variables* of type X.

(c) If X and Y are types, there is an associate set $Op_{X,Y}$. An element of this set is called an *operation symbol* of source X and of target Y. If $X = 1$, an operation symbol is called a constant of type Y.It is assumed that there are projection symbols $\pi_1 \in Op_{(X,Y),X}$ and $\pi_2 \in Op_{(X,Y),Y}$.

(d) For each type X there is an associate set Rel_X whose elements are called the *relation symbols* of signature X. We require at least the following relation symbols :

   - two relation symbols of signature 1, denoted $\top$ and $\bot$ and called respectively *true* and *false*;

   - for each type X, a relation symbol of signature (X, X), denoted by $=_X$, or simply $=$, and called *equality*;

   - for each type X a relation symbol of signature $(X, P(X))$ called the *membership relation* and denoted by $\in_X$.

(e) We also have the logical symbols : the conjunction symbol $\wedge$ and the collection or *abstraction symbol* $\{\dots \mid \dots\}$ which we read : the set of ... such that ....

8.1.2   Terms and formulas

If L is a language of higher order, we define by induction the (well-formed) expressions of L, being either *terms* or *formulas*, and we equip each term with a type.

(a) Each variable x of V_X is a term of type X.

(b) $\emptyset$ is a term of type 1.

(c) If t_1 and t_2 are terms of type X_1 and X_2, then (t_1, t_2) is a term of type (X_1, X_2)

(d) If f is in $Op_{(X,Y)}$ and t_1 is a term of type X then $f(t)$ is a term of type Y.

(e) If R is in Rel_X and t is a term of type X, then $R(t)$ is a formula. Similarly if t is a term of type Ω, then $\emptyset \in_1 t$ is a formula.

(f) If φ and ψ are formulas then $\varphi \wedge \psi$ is a formula.

(g) If φ is a formula and x is a variable of type X, then $\{x \mid \varphi\}$ is a term. Similarly $\{\emptyset \mid \varphi\}$ is a term of type Ω.

The usual logical connectors are introduced as abbreviations

1. $\phi \leftrightarrow \psi$ stands for

$$\{x_1 \mid \phi\} =_{\{\Omega\}} \{x_1 \mid \psi\}.$$

2. $\phi \to \psi$ stands for

$$\phi \leftrightarrow \phi \wedge \psi.$$

3. $\top$ stands for $\emptyset =_1 \emptyset$.

4. $\forall x \phi$ stands for

$$\{x \mid \phi\} = \{x \mid \top\}.$$

5. $\bot$ stands for

$$\forall \xi (\emptyset \in_1 \xi\}.$$

6. $\neg \phi$ stands for $\bot \leftarrow \phi$

7. $\phi \vee \psi$ stands for

$$\forall \xi ((\phi \to (\emptyset \in \xi)) \to \psi \to (\emptyset \in \xi) \to (\emptyset \in \xi)).$$

8. $\exists x \phi$ stands for

$$\forall \xi (\forall x (\phi \rightarrow (\emptyset \in \xi))) \rightarrow (\emptyset \in \xi).$$

We will not describe in detail which variables in a given expression are free variables or which bounded, but will point out that the status of x in $\forall x \varphi$ or in $\exists x \varphi$, is exactly that of x in $\{x \mid \varphi\}$.

We shall also omit here the description of the substitution, in an expression E, of terms of types $X_1, \ldots, X_n$ for variables of corresponding types.

8.1.3   Propositional calculus

In order to describe a theory in the language $\mathcal{L}$, we must now describe the rules of deduction and the axioms. A proof is then a finite sequence of formulas such that each is an axiom or results from the previous ones according to the rules of deduction.

The fundamental deduction rule is *modus ponens* : from formulas φ and $\varphi \rightarrow \psi$ we can deduce ψ.

The purely intuitionistic propositional calculus can be axiomatized in different ways. Here is for example a list of axioms (for every formula φ, ψ and η) :

1) $\varphi \rightarrow (\psi \rightarrow \varphi)$

2) $(\varphi \rightarrow \psi) \rightarrow ((\varphi \rightarrow (\psi \rightarrow \eta)) \rightarrow (\varphi \rightarrow \eta))$

3) $((\varphi \rightarrow \eta) \wedge (\psi \rightarrow \eta)) \rightarrow (\varphi \vee \psi \rightarrow \eta)$

4) $\varphi \wedge \psi \rightarrow \varphi$

5) $\varphi \wedge \psi \rightarrow \psi$

6) $\varphi \rightarrow (\psi \rightarrow (\varphi \wedge \psi))$

7) $\varphi \rightarrow (\varphi \vee \psi)$

8) $\psi \rightarrow (\varphi \vee \psi)$

9) $(\varphi \wedge \neg \varphi) \rightarrow \psi$

10) $(\varphi \to \neg\varphi) \to \neg\varphi$

11) $\bot \to \varphi$

12) $\top$

Here is a list of theorems expressed in terms of logical equivalence. They suggest the algebraic structure of the set of equivalence classes of formulas : this structure is no longer that of a boolean algebra as in the classical case. If we add a few simple rules of deduction (as for instance from $\varphi \leftrightarrow \psi$ we deduce $\varphi \to \psi$, or from $\varphi \wedge \psi$ we deduce φ) we get an axiomatic of intuitionistic propositional calculus.

1) $\varphi \wedge \varphi \longleftrightarrow \varphi$

2) $\varphi \vee \varphi \longleftrightarrow \varphi$

3) $(\varphi \wedge \psi) \wedge \eta \longleftrightarrow \varphi \wedge (\psi \wedge \eta)$

4) $(\varphi \vee \psi) \vee \eta \longleftrightarrow \varphi \vee (\psi \vee \eta)$

5) $\varphi \wedge \psi \longleftrightarrow \psi \wedge \varphi$

6) $\varphi \vee \psi \longleftrightarrow \psi \vee \varphi$

7) $\varphi \wedge (\psi \vee \eta) \longleftrightarrow (\varphi \wedge \psi) \vee (\varphi \wedge \eta)$

8) $\varphi \vee (\psi \wedge \eta) \longleftrightarrow (\varphi \vee \psi) \wedge (\varphi \vee \eta)$

9) $(\varphi \vee \psi) \wedge \psi \longleftrightarrow \psi$

10) $(\varphi \wedge \psi) \vee \psi \longleftrightarrow \psi$

11) $\varphi \wedge \bot \longleftrightarrow \bot$

12) $\varphi \vee \top \longleftrightarrow \top$

13) $\varphi \wedge \top \longleftrightarrow \varphi$

14) $\varphi \vee \bot \longleftrightarrow \varphi$

15) $\varphi \longleftrightarrow \varphi$

16) $(\varphi \to (\psi \wedge \eta)) \longleftrightarrow ((\varphi \to \psi) \wedge (\varphi \to \eta))$

17) $((\varphi \vee \psi) \to \eta) \longleftrightarrow ((\varphi \to \eta) \wedge (\psi \to \eta))$

18) $((\varphi \to \psi) \wedge \varphi) \longleftrightarrow \varphi \wedge \psi$

19) $((\varphi \to \psi) \wedge \psi) \longleftrightarrow \psi$

This list is obviously already redundant. If we now define negation by

$$\neg\varphi \longleftrightarrow (\varphi \to \bot)$$

it is interesting to display a few intuitionistic theorems concerning this negation :

20) $\neg(\varphi \vee \psi) \longleftrightarrow (\neg\varphi \wedge \neg\psi)$

21) $\neg\neg\neg\varphi \longleftrightarrow \neg\varphi$

22) $\varphi \to \neg\neg\varphi$

23) $(\neg\varphi \vee \neg\psi) \to \neg(\varphi \wedge \psi)$

24) $\neg\neg(\varphi \vee \neg\varphi)$

25) $(\varphi \vee \neg\varphi) \to (\neg\neg\varphi \longleftrightarrow \varphi)$

26) $\neg\neg(\varphi \to \psi) \longleftrightarrow \neg(\varphi \wedge \neg\psi)$

27) $\neg\neg(\varphi \to \psi) \longleftrightarrow (\varphi \to \neg\neg\psi)$

28) $(\varphi \vee \psi) \to \neg(\neg\varphi \wedge \neg\psi)$

29) $(\varphi \wedge \psi) \to \neg(\neg\varphi \vee \neg\psi)$

30) $(\varphi \to \psi) \to \neg(\varphi \wedge \neg\psi)$

To conclude, here are a few examples of formulas that are theorems in the classical propositional calculus but not in intuitionistic calculus. Of course this list cannot be exhaustive but it contains the essential. We write $\nvdash$ to denote *non-theoremhood*.

(a) $\nvdash \varphi \vee \neg\varphi$;

(b) $\nvdash \neg\neg\varphi \rightarrow \varphi$
(but we have (21) and (22));

(c) $\nvdash \neg(\neg\varphi \wedge \neg\psi) \rightarrow (\varphi \vee \psi)$
(but we have (28)). We cannot in fact define $\vee$ from $\wedge$ and $\neg$.

(d) $\nvdash \neg(\neg\varphi \vee \neg\psi) \rightarrow (\varphi \wedge \psi)$
(but we have (29)). We cannot define $\wedge$ from $\vee$ and $\neg$.

(e) $\nvdash \neg(\varphi \wedge \neg\psi) \rightarrow (\varphi \rightarrow \psi)$
(but we have (30)). We cannot define the implication from $\wedge$ and $\neg$.

A practical way to test potential theorems can be described as follows. In classical propositional calculus the equivalence classes of formulas form a Boolean algebra. In fact, if X is any set, $\mathcal{P}(X)$, the set of subsets of X, is a Boolean algebra. In order to find out whether a formula is tautological from the propositional point of view, it suffices to translate in terms of subsets of X, interpreting $\wedge$ as $\cap$, $\vee$ as $\cup$, $\neg$ as the complement. In intuitionistic propositional calculus, the equivalence classes of formulas form a *Heyting algebra*. In fact, if E is a topological space, $Open(E)$, the set of open subsets of E, is a Heyting algebra. We interpret then $\wedge$ as $\cap$, $\vee$ as $\cup$, $\top$ as the entire space, $\perp$ as the empty space, $\neg$ as the interior of the complement :

$$\neg U = Int(C_E U)$$

and $\rightarrow$ with

$$(U \rightarrow V) = Int(C_E U \cup V).$$

For instance, by taking $E = \mathbb{R}$ and $U = \{\alpha \in \mathbb{R} \mid \alpha > 0\}$, we see that $\neg U = \{\alpha \in \mathbb{R} \mid \alpha < 0\}$.

We can then verify that a formula of propositional calculus is an intuitionistic tautology if and only if it becomes a true statement when interpreted in any topological space E .

In the example quoted above where $U = \{\alpha \in \mathbb{R} \mid \alpha > 0\}$ and $\neg U = \{\alpha \in \mathbb{R} \mid \alpha < 0\}$, we see that $U \cup \neg U = \mathbb{R}\backslash\{0\}$. Thus it is not the entire space $\mathbb{R}$. This shows that $\varphi \vee \neg\varphi$ cannot be an intuitionistic tautology.

The axioms concerning the equalities (of any type) are the usual axioms :

1) $x =_X x$ where x is a variable of type X.

2) $x =_X y \to y =_X x$ where x and y are variables of type X.

3) $(x =_X y \wedge y =_X z) \to x =_X z$ where x, y and z are variables of type X.

4) If f is an operation of source $X = (X_1, \ldots, X_n)$ and of target Y,

$$(x_1 =_{X_1} x_1') \wedge \ldots \wedge (x_n =_{X_n} x_n') \to f(x_1, \ldots, x_n) =_Y f(x_1', \ldots, x_n')$$

where x_i and x_i' are of type X_i.

5) If R is a relation of signature $X = (X_1, \ldots, X_n)$ we have
$$((x_1 =_{X_1} x_1') \wedge \ldots \wedge (x_n =_{X_n} x_n') \wedge R(x_1, \ldots, x_n)) \to R(x_1', \ldots, x_n')$$
where the x_i and x_i' are of type X_i.

8.1.4 Predicate calculus

We now present the fundamental rules and axioms of *predicate calculus*.

1) From $\varphi \to \psi$ we can deduce

$$\varphi \to \forall x \psi(x)$$

if x is not a free variable of φ.

2) From $\varphi \to \psi$ we can deduce

$$\exists x \varphi(x) \to \psi$$

if x is not a free variable of ψ.

3) If φ is a formula, x a variable of type X and t a term of type X, we have
$$\forall x \varphi(x) \to \varphi(t).$$

4) If φ is a formula, x a variable of type X and t a term of type X, we have
$$\varphi(t) \to \exists x \varphi(x).$$

The following substitution rules are easily inferred from the preceding rules and axioms :

5) If x is a variable of type X and t a term of type X and if the free variables of t are not bounded in φ then if we have φ we do also have

$\varphi(t \mid x)$, where $\varphi(t \mid x)$ results from the substitution of t for x. This formula is often denoted by $\varphi(t)$.

In classical predicate calculus one generally refers either to (1) and (3) or to (2) and (4) because classically we have interdefinability

$$\forall x \longleftrightarrow \neg \exists x \neg \quad \text{and} \quad \exists x \longleftrightarrow \neg \forall x \neg.$$

In the intuitionistic case, we have

6)
$$\exists x \varphi \to \neg \forall x \neg \varphi$$

but
$$\nvdash \neg \forall x \neg \varphi \to \exists x \varphi$$

and

7)
$$\forall x \varphi \to \neg \exists x \neg \varphi$$

but
$$\nvdash \neg \exists x \neg \varphi \to \forall x \varphi.$$

Let us also quote two theorems of intuitionistic predicate calculus :

8)
$$\vdash \forall x (\varphi \to \psi(x)) \longleftrightarrow (\varphi \to \forall x \psi(x))$$

if x is not free in φ.

9)
$$\vdash \forall x (\varphi(x) \to \psi) \longleftrightarrow (\exists x \varphi(x) \to \psi)$$

if x is not free in ψ. On the other hand we have only

10)
$$\vdash \exists x (\varphi(x) \to \psi(x)) \to (\forall x \varphi(x) \to \exists x \psi(x))$$

but
$$\nvdash (\forall x \varphi(x) \to \exists x \psi(x)) \to \exists x (\varphi(x) \to \psi(x)).$$

These observations show that in intuitionistic logic it is difficult to prove an existential formula. The only quoted rule that allows us to derive an existential formula is Rule 4. This suggests that the intuitionistic way to show an existential formula $\exists x \varphi(x)$ is to construct explicitly a term t such that we have $\varphi(t)$.

8.1.5   Higher order theory

It remains only to see what is specific to the "set" or "higher order" aspects of the situation.

1) If φ is a formula, $(x_1, \ldots, x_n)$ are variables of types $(X_1, \ldots, X_n)$ and $(t_1, \ldots, t_n)$ are terms of type $(X_1, \ldots, X_n)$, we have

$$\varphi(t_1, \ldots, t_n) \longleftrightarrow (t_1, \ldots, t_n) \in_X \{x_1, \ldots, x_n \mid \varphi\}.$$

Notice that for $n = 0$ this scheme includes

$$\varphi \longleftrightarrow \emptyset \in_\emptyset \{\emptyset \mid \varphi\}.$$

2) The second axiom is that of *extensionality*. Let $(x_1, \ldots, x_n)$ be variables of types $(X_1, \ldots, X_n)$ and let t and s be two terms of type $P(X_1, \ldots, X_n)$. Denote $(X_1, \ldots, X_n)$ by X. If

$$((x_1, \ldots, x_n) \in_X t) \longleftrightarrow ((x_1, \ldots, x_n) \in_X s)$$

then

$$s =_{P(X)} t.$$

In particular for $n = 0$, if

$$\emptyset \in_\emptyset t \longleftrightarrow \emptyset \in s$$

then

$$s =_\Omega t.$$

Remark. We can see the object $\Omega(= P(\emptyset))$ as the *truth values object* of the theory . From this point of view, along with the formula $\top$ (which is in $Rel_\emptyset$), we can also consider the term

$$\{\emptyset \mid \top\}$$

which is a term of type Ω that we can call the true one. Notice that since $\top$ is a theorem, and since we have Axiom 1

$$\top \longleftrightarrow \emptyset \in_\emptyset \{\emptyset \mid \top\},$$

we also have

$$\emptyset \in_\emptyset \{\emptyset \mid \top\}.$$

Thus φ is a theorem if and only if

$$\{\emptyset \mid \varphi\} = \{\emptyset \mid \top\}.$$

8.2   The concept of topos

8.2.1   (Insufficient) introduction to categories

When mathematicians are studying a given structure, they are concerned not only with the class of objects equipped with that structure but also with the class of maps adapted to that structure. Topologists are not only interested in topological spaces but also, and perhaps more importantly, in continuous functions. Linear operators are more important for algebraists than vector spaces.

At the same time, however, the class of maps for a given structure (continuous, linear, differentiable,...) itself has an algebraic structure. This algebraic structure is the main motivation for the concept of category which we now define.

Definition 1. *A category comprises :*

1. *two classes C_o and C_1 (the class of objects and the class of arrows or morphisms)*

2. *two maps $\alpha : C_1 \to C_o$ and $\beta : C_1 \to C_o$ (which to an arrow $f \in C_1$ associates its source and its target. If $\alpha(f) = X$ and $\beta(f) = Y$ we write simply $f : X \to Y$) ;*

3. *a partially defined operation on C_1. If we have $X \xrightarrow{f} Y \xrightarrow{g} Z$ (and so $\beta(f) = \alpha(g)$), we have a composite $X \xrightarrow{g \circ f} Z$;*

4. *a map $\epsilon : C_o \to C_1$ (if X is an object, $\epsilon(X)$ is said to be the identity on X and often denoted by 1_X.*

These data must satisfy the following axioms:

1. *Sources and target axioms:*

    $$\alpha(\epsilon(X)) = X \, , \ \beta(\epsilon(X)) = X \, , \ \alpha(g \circ f) = \alpha(f) \, , \ \beta(g \circ f) = \beta(g);$$

2. *associativity axiom: $h \circ (g \circ f) = (h \circ g) \circ f$ (when it makes sense) ;*

3. *identity axioms : if $X \xrightarrow{f} Y$ then*

    $$f \circ \epsilon(X) = f = \epsilon(Y) \circ f.$$

Examples

1) The fundamental examples correspond to various mathematical theories. If C_o is the class of sets (resp. groups, rings, topological spaces, vector spaces, manifolds,...) and C_1 the class of maps (resp. group homomorphisms, ring homomorphisms, continuous maps, linear maps, differentiable maps, ...), we obtain a category with the usual notions of source, target, composition and identity.

2) A monoid is a category with only one object, its identity element. In particular a group is a category with one object.

3) Let $(X, \leq)$ be a partially ordered set. We can define a corresponding category with $C_o = X, C_1 = \{(a,b)|a \leq b\}, \alpha(a,b) = a, \beta(a,b) = b, (b,c) \circ (a,b) = (a,c), \epsilon(a) = (a,a)$. In particular, if E is a topological space $\underline{Open}(E)$ is a category whose objects are open subsets in E and arrows can be seen as canonical injections of one open subset into another.

Definition 2. Let $C = (C_o, C_1)$ and $D = (D_o, D_1)$ be two categories. A functor $F : C \to D$ is a pair of maps $T : C_o \to D_o$ and $T : C_1 \to D_1$ preserving the structure of category, i.e.: $T(\alpha(f)) = \alpha(T(f)), T(\beta(f)) = \beta(T(f)), T(\epsilon(X)) = \epsilon(T(X))$ and, if $g \circ f$ is defined, $T(g \circ f) = T(g) \circ T(f)$.

Intuitively, if we think of C and D as "mathematical theories", we can think of functors as machines to translate concepts of the first theory into concepts of the second theory.

We would like to speak about a category of all categories and functors between them. This however gives rise to a set-theoretical problem - just as when we try to talk of the set of all sets. Thus we call a category $C = (C_o, C_1)$ *small* if C_o and C_1 are sets. It is clear that $Cat = (Cat_o, Cat_1)$ - where Cat_o is the class of small categories and Cat_1 the class of functors between small categories - is a category. Here we use a naive distinction between sets and classes, the later being "illegitimate big sets", in fact, actually, these classes are like "not collectivising predicates" (in Bourbaki's terminology).

It is interesting to observe that if C is a category we can obtain a new category C^o (the opposite or the dual of C) just interchanging source and target or, in other words, formally reversing the arrows. A functor from C^o to D is then said to be a *contravariant functor* from

C to *D*.

Categories were originally define to allows a definition of the concept of naturality. If we think two functors F and G from C to D as being two translation machines , we must have a concept of comparison between translations.

Definition 3. *Let*

$$C \underset{G}{\overset{F}{\rightrightarrows}} B$$

be two functors. A natural transformation $\phi : F \Rightarrow G$ *is a map from* C_0 *to* D_1 *associating to an object* X *an arrow* $\phi_X : FX \to GX$ *in* D_1 *such that for all* $f : X \to Y$ *in* C_1, *the following square commutes*

$$
\begin{array}{ccc}
FX & \xrightarrow{\;Ff\;} & FY \\
\phi_X \downarrow & & \downarrow \phi_Y \\
GX & \xrightarrow[Gf]{} & GY
\end{array}
$$

i.e. $Gf \circ \phi_X = \phi_Y \circ Ff$.

We shall always suppose that between two objects X and Y in a category C there is only a set of arrows. With that assumption we can describe for all X in C a functor (very useful)

$$h^X : C \to Set$$

by $h^X(Y) = C(X, Y)$, i.e. the set of arrows in C from X to Y ; and for $v : Y \to Y'$ we define $h^X(v)$ by composition:

$$h^X(v) : C(X, Y) \to C(X, Y') : f \mapsto v \circ f.$$

In the same way we can define a contravariant functor

$$h_X : C \to Set$$

by $h_Y(X) = C(X, Y)$ i.e. the set of arrows in C from X to Y ; and for $u : X \to X''$ we define $h_Y(u)$ by composition:

$$h_Y(u) : C(X', Y) \to C(X, Y) : f \mapsto f \circ u.$$

If $u : X \to X'$ we construct the natural transformation $h^u : h^{X'} \Rightarrow h^X$ by $(h^u)_Y(g) = g \circ u$ where $g \in C(X', X)$.

Thus we have constructed a contravariant functor from the category C to the category Set^C whose objects are functors from C to Set, and morphisms are natural transformations between such functors.

Definition 4. *A functor F from C to Set is said to be representable if it is naturally isomorphic to a functor of the form h^X for some object X. We then say that X represents F.*

In some sense, we enter more deeply into a mathematical concept if we can express it "with arrows". Here are some simple examples of concepts expressible with arrows.

a) An arrow $u : A \to B$ is a *monomorphism* if for all f and g, $u \circ f = u \circ g$ implies $f = g$. It is easy to see that in some 'concrete" categories such that categories of sets, of groups, of topological spaces or of R-modules (for a ring R), the monomorphisms are just the injections. But this is not always the case ; it turns out that the concept of monomorphism is somehow deeper than that of injection.

b) An arrow $v : A \to B$ is an *epimorphism* if for all f and g, $f \circ v = g \circ v$ implies $f = g$. It is easy to seee that in the category of sets or that of R-modules, the epimorphisms are just the surjections. It is not so easy to see - but it is true - that the same goes for the category of groups. In the category of rings and in the category of Hausdorff topological spaces there exist non-surjective epimorphisms.

c) An *isomorphism* $i : A \to B$ is an invertible arrow. Thus i is an isomorphism if there exists $j : B \to A$ such that $i \circ j = 1_A$ and $j \circ i = 1_B$. In many concrete categories isomorphisms are the bijective morphisms but this is not always the case. For exemple in the category of Hausdorff topological spaces there are bijective continuous maps which are not isomorphisms.

d) If A and B are two objects in C we say that

$$A \xleftarrow{\pi_1} C \xrightarrow{\pi_2} B$$

is a *product* of A and B if this diagram has a kind of universality in the sense that for any diagram of the same shape

$$A \xleftarrow{f_1} X \xrightarrow{f_2} B$$

there exists one and only one arrow $h : X \to C$ such that $f_1 = \pi_1 \circ h$ and $f_2 = \pi_2 \circ h$. If such a product exists, it is unique up to a unique isomorphism and thus we allow ourselves to speak of "the product" of A and B, and to write $A \times B$.

e) A *terminal object* $\underline{1}$ in a category C is an object such that for every object A there is a unique arrow from A to $\underline{1}$. For sets, groups, topological spaces, R-modules or manifolds, the terminal object is a singleton, endowed with the appropriate (trivial) structure.

f) Let $A \underset{g}{\overset{f}{\rightrightarrows}} B$ be a pair of arrow in C. An *equalizer* of f and g is an object K with an arrow $k : K \to A$ equalizing f and g (i.e. such that $f \circ k = g \circ k$) in a universal manner in that sense that if $u : X \to A$ equalizes f and g there exists a unique facorization of u through k, say $u = k \circ u'$. In sets, groups, topological spaces, R-modules, equalizers exist. But in the category of differentiable manifolds two parallel arrows f and g have an equalizer only if they satisfy a transversality condition.

Each concept defined with arrows has a dual defined by reversing the arrows. For example an arrow u is a monomorphism in C if and only if it is an epimorphism in C^o. In the same spirit we can define the dual of the product, called the *coproduct* or the sum, the dual of the terminal object, called the *initial object*, or the dual of an equalizer, called a *coequalizer*.

Defintion 5. *A category C is finitely complete if the product of two objects, the terminal object, and the equalizer of any pair of parallel arrows exist in C.*

In a finitely complete category other constructions can be made. As only one useful example we give that of pullbacks. We consider two arrows with a common target X. We say that the square

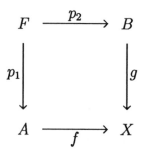

is a *pullback* if it commutes in a universal way, i.e. it commutes, and any square over f and g ($f \circ z_1 = g \circ z_2$) factors uniquely throught F. For example in sets we can take $F = \{(a, b) | f(a) = g(b)\}$.

Proposition 1. *In a finitely complete category pullbacks always exist.*

Proof: If we have $A \xleftarrow{f} X \xrightarrow{g} B$ we take the equalizer k of $f \circ \pi_1$ and $g \circ \pi_2$:

$$F \xrightarrow{k} A \times B \underset{g \circ \pi_2}{\overset{f \circ \pi_1}{\rightrightarrows}} X$$

It is easy to prove that F is the pullback of f and g with $p_1 = \pi_1 \circ k$ and $p_2 = \pi_2 \circ k$ as projections. ∎

We finish these trivialities with the notion of subobject. For an object X there are in general many monomorphisms with X as target. We say two such monomorphisms are equivalent if they differ only by an isomorphism of their sources. An equivalence class of monomorphisms with target X is called a *subobject* of X.

Definition 6. *A category C is said to be well powered if, for any X, there is only a set of subobjects of X. This set of subobjects of X is denoted by $\mathcal{P}(X)$.*

If $X_1 \xrightarrow{f} X_2$ is an arrow in a well powered and finitely complete category C we describe a map

$$\mathcal{P}(X_2) \xrightarrow{\mathcal{P}(!)} \mathcal{P}(X_1)$$

(which can also be denote by f^{-1}). Let $u : A \to X_2$ be a monomorphism representing a subobject $\underline{A}$ of X_2. We consider the pullback of f an u :

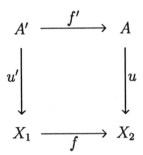

It is easy to verify that u' is also a monomorphism. The equivalence class $\underline{A}'$ determined by u' is $\mathcal{P}(f)(A)$ (or $f^{-1}(A)$).

Trivially, if $f = 1_X, \mathcal{P}(1_X) = 1_{\mathcal{P}(X)}$. If we have $X_1 \xrightarrow{f} X_2 \xrightarrow{g} X_3$,

$$\mathcal{P}(g \circ f) = \mathcal{P}(f) \circ \mathcal{P}(g).$$

Thus $\mathcal{P}$ is a contravariant functor from C into Set.

8.2.2 Definition and examples of toposes

Every topos is a finitely complete well powered category. But this is "general abstract nonsense" common to almost all mathematical theories thought of as categories, and we require much more of a topos.

The fundamental and original point is that in a topos we can mimic widely what happens with sets, thanks to the existence of a concept of "object of subobjects" in the category. This "object of subobjects" of an object will allow us to talk about relations in the category in the same way that $P(X)$ allow us to talk about relations in the category of sets. (Of course $P(X)$, or $\mathcal{P}(X)$, is the set of subsets of X).

For sets a relation R between Y and X is a subset $R \subseteq Y \times X$; this is a set of pairs. But R determines a map

$$\phi_R : Y \to P(X)$$

defined by

$$\phi_R(y) = \{x \in X | R(y, x)\}.$$

Conversly, a map $\phi : Y \to P(X)$ corresponds to a relation R_ϕ described by

$$R_\phi(y, x) \text{ iff } x \in \phi(y).$$

In short, for sets we have a canonical bijection

$$P(Y \times X) \simeq Set(Y, P(X))$$

(where $Set(Y, Z)$ is the set of maps from Y to Z.

For sets the preceding bijection is "natural" in Y in the sense that for all $g : Y' \to Y$ the following square commutes :

$$P(Y \times X) \xrightarrow{\ \simeq\ } Set(Y, P(X))$$

$$\Big\downarrow{g^*} \qquad\qquad \Big\downarrow{g^o}$$

$$P(Y' \times X) \rightleftarrows Set(Y', P(X))$$

where, for $R \in P(Y \times X)$, we put

$$g^*(R)(y', x) \text{ iff } R(g(y'), x)$$

and, for $\phi : Y \to P(X)$, we put

$$g^o(\phi) = \phi \circ g.$$

Now we return to a well powered and finitely complete category $\mathcal{E}$. If X and Y are objects in $\mathcal{E}$, we take $P(X \times Y)$ as the set of relations from Y to X.

We suppose now that, for each object X in $\mathcal{E}$, we have an object, denoted by $P(X)$ and a bijection between $P(Y \times X)$ and $\mathcal{E}(Y, P(X))$ (the set of arrows from Y to $P(X)$ and we suppose that this bijection is natural. This means that for all $g : Y' \to Y$ in $\mathcal{E}$, the following square commutes :

$$P(Y \times X) \xrightarrow{\ \simeq\ } \mathcal{E}(Y, P(X))$$

$$\Big\downarrow{g^*} \qquad\qquad \Big\downarrow{g^o}$$

$$P(Y' \times X) \rightleftarrows \mathcal{E}(Y', P(X))$$

where g^* is defined by pullback and g^o by composition.

Definition 7. *Let $\mathcal{E}$ be a well-powered, finitely complete category. We say that relations are representable if, for all X in $\mathcal{E}$, there is an object $P(X)$ and for all Y an isomorphism*

$$\mathcal{P}(Y \times X) \simeq \mathcal{E}(Y, P(X))$$

natural in Y.

Definition 8. *A topos is a well powered, finitely complete category in which relations are representable.*

Examples

I.- The first example of topos is of course the category *Set* of sets and maps.

II.- As a second example we will consider the category of *varying sets* .

Let $\underline{2}$ be the category $0 \to 1$, and $\mathcal{E} = \underline{Set}^2$ the category of functors and natural transformations from $\underline{2}$ to *Set*. For the beginner we explicit the intuitive content of this category

a) A varying set $\underline{X}$ is just a map $X_0 \xrightarrow{f} X_1$. We think of X_0 as what $\underline{X}$ was "before", X_1 as what $\underline{X}$ is "after" and f as an act of transformation. An object is $X_0 \xrightarrow{f} X_1$ as a single entity.

b) A varying map $\underline{\phi} : \underline{X} \to \underline{Y}$ is a commutative square (ϕ_0, ϕ_1) :

$$
\begin{array}{ccc}
X_0 & \xrightarrow{\ \phi_0\ } & Y_0 \\
\downarrow{\scriptstyle f} & & \downarrow{\scriptstyle g} \\
X_1 & \xrightarrow[\ \phi_1\ ]{} & Y_1
\end{array}
$$

c) We define composition

$$\underline{X} \xrightarrow{\ \phi\ } \underline{Y} \xrightarrow{\ \psi\ } \underline{Z}$$

by

$$(\psi_0, \psi_1) \circ (\phi_0, \phi_1) = (\psi_0 \circ \phi_0, \psi_1 \circ \phi_1)$$

and the identity on $\underline{X}$ by $(1_{X_0}, 1_{X_1})$. So we have the category $\underline{Set^2}$ of varying sets.

Proposition 2. *The category* $\mathcal{E} = \underline{Set^2}$ *of varying sets is a topos.*

Proof: The product $\underline{X} \times \underline{Y}$ of two varying sets

$$\underline{X} \xleftarrow{\pi_1} \underline{X} \times \underline{Y} \xrightarrow{\pi_2} \underline{Y}$$

is described by

$$
\begin{array}{ccccc}
X_0 & \xleftarrow{\pi_{10}} & X_0 \times Y_0 & \xrightarrow{\pi_{20}} & Y_0 \\
\downarrow{\scriptstyle f} & & \downarrow{\scriptstyle f \times g} & & Sarg \\
X_1 & \xleftarrow{\pi_{11}} & X_1 \times Y_1 & \xrightarrow{\pi_{21}} & Y_1
\end{array}
$$

The terminal object $\underline{1}$ is $\{*\} \xrightarrow{1_{\{*\}}} \{*\}$. In fact for any $\underline{X}$ there is one and only one varying map from $\underline{X}$ to $\underline{1}$:

$$
\begin{array}{ccc}
X_0 & \longrightarrow & \{*\} \\
\downarrow{\scriptstyle f} & & Sar1_{\{*\}} \\
X_1 & \longrightarrow & \{*\}
\end{array}
$$

A varying subobject $\underline{A}$ of $\underline{X}$ can be described by $A_0 \subseteq X_0$ and $A_1 \subseteq X_1$ such that $f(A_0) \subseteq A_1$. Thus a varying relation $\underline{R}$ from $\underline{X}$ to $\underline{Y}$ is $R_0 \subseteq X_0 \times Y_0$ and $R_1 \subseteq X_1 \times X_1$ such that

$$R_0(x_0, y_0) \text{ implies } R_1(f(x_0), g(y_0)).$$

Let $X \overset{\phi}{\underset{\psi}{\rightrightarrows}} Y$ be a pair of varying maps. The equalizer of ϕ and ψ is given by the subobject $\underline{K}$ of $\underline{X}$ where $K_0 \subseteq X_0$ is the set $\{x_0 \in X_0 | \phi_0(\pi_0) = \psi_0(x_0)\}$ and $K_1 \subseteq X_1$ is the set $\{x_1 \in X_1 | \phi_1(\pi_1) = \psi_1(x_1)\}$

Trivially we have a set $\mathcal{P}(\underline{X})$ of varying subobjects of a varying set $\underline{X}$. But we will see that we have also a varying set $\underline{P(X)}$ which represents varying relations. We take

$$\underline{P(X)} = (P_0(\underline{X}) \overset{p}{\longrightarrow} P_1(\underline{X}))$$

where $P_0(\underline{X})$ is the set $\mathcal{P}(\underline{X})$ of varying subobjects of $\underline{X}$, $P_1(\underline{X})$ is the set $\mathcal{P}(X_1)$ of subsets of X_1 and $p(A_0, A_1) = A_1$. We must show that $\underline{P}$ represents relations.

We consider a varying relation $\underline{R} \subseteq \underline{Y} \times \underline{X}$. We associate to $\underline{R}$ the varying map (r_0, r_1)

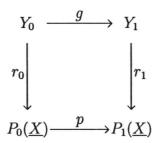

given by

$$r_0(y_0) = (B_0, B_1)$$

with

$$B_0 = \{x_0 \in X_0 | R_0(y_0, x_0)\}, B_1 = \{x_1 \in X_1 | R_1(g(y_0), x_1)\}$$

Conversely to $(r_0, r_1) : \underline{Y} \to \underline{P(X)}$ we associate $\underline{R} = (R_0, R_1)$ given by

$$R_0(y_0, x_0) \text{ iff } x_0 \in B_0 = r_0(y_0)_0$$

$$R_1(y_1, x_1) \text{ iff } x_1 \in r_1(y_1).$$

So we have constructed a bijection between the set $Rel(\underline{Y}, \underline{X}) = \mathcal{P}(\underline{Y} \times \underline{X}$ and the set $E(\underline{Y}, \underline{P(X)})$ of varying maps. It is easy to show that this construction is natural. ∎

III.- In the preceding example, sets vary over a kind of time with two states 0 and 1 and only one transition. We obviously can generalize to a "time" represented by an ordered or preordered set. But we can also accept many different transitions between two instants. Such a generalized time is just a small category.

Thus we consider the category $\mathcal{E} = \underline{Set}^C$ of functors and natural transformations from a small category C to the category of sets. We can see an object $E : C \to \underline{Set}$ as a varying set. We suggest the desired intuition by saying that an object A of C is a *stage* and that an arrow $u : A \to B$ in C is a transition between stages A and B. The varying set $E : C \to \underline{Set}$ is viewed as being defined by its values at the different stages A and its evolutions $E(u) : E(A) \to E(B)$ for all A and all u ; the coherence of the evolutions is expressed by the fact that E is a functor. Varying maps are natural transformations.

The category $\mathcal{E}$ is finitely complete. The product $E \times F$ is given at stage A by the product of sets $E(A) \times F(A)$. The equalizer $k : K \to E$ of two arrows $E \overset{\phi}{\underset{\psi}{\rightrightarrows}} F$ at stage A is the equalizer of ϕ_A and ψ_A, i.e. the set of elements x of $E(A)$ such that $\phi_A(x) = \psi_A(x)$. The terminal object of $\mathcal{E}$ is the "constant singleton", i.e. the functor 1 given by $1(A) = \{*\}$ for all A (and, of course, $1(u)$ is the identity for all u). As C is small, there is only a set of arrows with target an object E of $\mathcal{E}$, thus $\mathcal{E}$ is well-powered.

Proposition 3. *The category $\mathcal{E} = \underline{Set}^C$, where C is a small category, is a topos.*

Proof: It remains to prove that relations are representable. We shall only describe the varying set $P(E)$ for $E : C \to \underline{Set}$.

a) For X in C we consider the functor $h^X : C \to \underline{Set} : Y \to \mathcal{E}(X, Y)$. Then we put

$$P(E)(X) = \mathcal{P}(h^X \times E).$$

So $P(E)(X)$ is viewed as the set of subfunctors of $h^X \times E$, or as the set of relations from h^X to E.

b) For $u : X \to X'$ in C we must define $P(E)(u) : P(E)(X) \to P(E)(X')$, which is done by pullback : if $G \to h^X \times F$ is a subfunctor i.e. an element of $P(E)(X)$ we define $G' = P(E)(u)(G)$ by the pullback

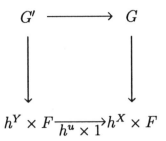

in $\underline{Set}^C$. ∎

IV.- For every small category C, the category $\underline{Set}^{C^o}$ of contravariant functors from C to $\underline{Set}$ is a topos. No new proof is necessary. This category is called the category of *presheaves* on C.

V.- We will now describe the case of *sheaves* on a topological space. Let X be a topological space and $\underline{Open}(X)$ the category (associated to the partially ordered set) of open subsets of X (cf. 8.2.1. example (3)).

A presheaf on X is a varying set over $\underline{Open}(X)^o$ i.e. a contravariant functor from $\underline{Open}(X)$ to sets. Let F be such a presheaf. If $U' \subseteq U$ is an inclusion of open subsets, the corresponding map $\rho_{U,U'} : F(U) \to F(U')$ is called a restriction. That name is coming from the paradigmatic case where $F(U)$ is a set of (continuous) maps from U to a fixed topological space (for example $\mathbb{R}$) and $\rho_{U,U'}$ is actually "restriction to U' " .

It is interesting to consider a subcategory of the category of presheaves (or varying sets) whose objects satisfy a compatibility condition with the topology of X. A sheaf is a kind of "continuously" varying set a point of view first proposed by F.W. Lawvere . More precisely

Definition 9. *A presheaf F on a topological space X is called a sheaf if it satisfies the following coherence condition (C). Let U be an open subset of X and $(U_i)_{i \in I}$ a family of open subsets such that $U = \bigcup_{i \in I} U_i$. We consider for each i in I an element $s_i \in F(U_i)$. The condition (C) is : If*

$$(\forall i \in I) \, (\forall j \in I) \, \rho_{U_i, U_i \cap U_j}(s_i) = \rho_{U_j, U_i \cap U_j}(s_j)$$

then there exists one and only one s in $F(U)$ such that, for all i in I,

$$\rho_{U,U_i}(s) = s_i.$$

Let F and G be two sheaves over X. A morphism $\phi : F \to G$ is a map of varying sets, i.e. for each U open in X a map $\phi_U : F(U) \to G(U)$

in such a manner that it is natural (in other words compatible with restriction). Products of two sheaves, the terminal sheaf, and equalizers are built as for varying sets. For example we easily verify that if F and G are sheaves, the presheaf given by $(F \times G)(U) = F(U) \times G(U)$ is in fact a sheaf.

A subsheaf A of a sheaf F is a sheaf such that for all U in $Open(X)$, $A(U) \subseteq F(U)$. For example a sheaf relation from F to G is a subsheaf of $F \times G$. We have thus the set $\mathcal{P}(F)$ of subsheaves of F and the category $Sh(X)$ of sheaves on X is well-powered.

Proposition 4. *The category $Sh(X)$ of sheaves over a topological space X is a topos.*

Proof: Let F be a sheaf over X. If U is open in X, we consider $F_{|_U}$ as the sheaf over U given for each open subset U' in U by

$$F_{|_U}(U') = F(U')$$

We denote by $P(F)(U)$ the set of subsheaves of $F_{|_U}$. Obviously, if $G \in P(F)(U)$ (G is a subsheaf of $F_{|_U}$) and if $U' \subseteq U, then G_{|_{U'}}$ is a subsheaf of $F_{|_{U'}}$. So we have a restriction map

$$\rho_{U,U'} : PF(U) \to PF(U') : G \to G_{|_{U'}}$$

and it is easy to verify that the varying set $P(F)$ so defined is in fact a sheaf.

Now we consider two sheaves F and G. We must show that the set $\mathcal{P}(G \times F)$ of sheaf relations from G to F is in bijection with the set $\mathcal{E}(G, PF)$ of morphisms from G to PF and that these bijections are natural in G. We only describe the two maps

$$\mathcal{P}(G \times F) \rightleftharpoons \mathcal{E}(G, PF)$$

and leave the verifications to the reader. To a sheaf relation $R \subseteq G \times F$ we associate $\phi_R : G \to PF$ by

$$\phi_R(U) : G(U) \to PF(U) : s \to \phi_R(U)(s)$$

where $\phi_R(U)(s)$ is the subsheaf of $F_{|_U}$ given, for $U' \subseteq U$, by

$$\phi_R(U)(s)(U') = \{t \in F(U')|(\rho_{U,U'}(s), t) \in R(U')\}.$$

Conversely to $\phi : G \to PF$ we associate the sheaf relation $R_\phi \subseteq G \times F$ given by

$$R_\phi(U) = \{(s,t) \in G(U) \times F(U) | t \in \phi_U(s)(U)\}.$$

■

VI.- It is possible, as observed by Grothendieck, to generalize the concept of sheaf to other categories than a category of open subsets. Let C be a small category. For X an object in C we consider the contravariant functor

$$h_X : C \to \underline{Set} : Y \to C(Y, X)$$

which can be thought of as the set of all arrows with target X. A *sieve* on X is a subfunctor of h_X. We can think of a sieve as a set of arrows with target X, saturated under left composition.

Definition 10. *A localizing system $\mathcal{R}$ on C consists of a set $\mathcal{R}(X)$ of sieves on X, for each object X, such that if the square*

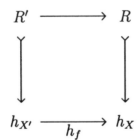

is a pullback with R in $\mathcal{R}(X)$ then R' is in $\mathcal{R}(X')$.

Grothendieck's idea was that a sieve R in $\mathcal{R}(X)$ is a kind of generalization of an open covering in topological spaces.

Definition 11. *Let C be a small category and $\mathcal{R}$ a localizing system on C. We consider a presheaf $F : C^o \to \underline{Set}$. This presheaf is called a sheaf on $((C, \mathcal{R})$ if for each X in C and each $R \to h_X$ in $\mathcal{R}(X)$, every natural transformation $\alpha : R \Rightarrow F$ extends uniquely to a natural transformation $\overline{\alpha} : h_X \Rightarrow F$.*

A natural transformation $\beta : h_X \to F$ is an element of F at stage X - explicitly the element $\beta_X(id_X)$ (which in fact is in $F(X)$). We can see a natural transformation $R \Rightarrow F$ as a partial (or local) element given only on R. The condition of Definition 11 says that any element given locally on R glues together into an element of F at stage X. A morphism of sheaves is a morphism of presheaves and thus a natural tranformation. So we obtain the category $Sh(C, \mathcal{R})$ of sheaves on $(C, \mathcal{R})$.

We assume without proof the following proposition

Proposition 5. *If C is a small category and $\mathcal{R}$ is a localizing system on C, $Sh(C, \mathcal{R})$ is a topos.*   ∎

For a proof see for example, F. Borceux [5] "Handbook of categorical algebra" vol.3.

8.2.3   Toposes and set theoretical notions

Inside a topos there is a logic. We begin with some set-theoretical notions. If $\mathcal{E}$ is a topos there is a bijection

$$\mathcal{P}(X \times Y) \simeq E(Y, PX)$$

(because $X \times Y \simeq Y \times X$). When $Y = PX$ the relation corresponding to the arrow 1_{PX} is called the *membership relation* and is written

$$\in_X \to X \times PX.$$

Proposition 6. *Let $R \to X \times PX$ be a relation. The corresponding morphism $\phi_R : Y \to PX$ is the unique morphism such that we have a pullback*

$$
\begin{array}{ccc}
R & \longrightarrow & \in_X \\
\downarrow & & \downarrow \\
X \times Y & \xrightarrow[i_X \times \phi_R]{} & X \times PX
\end{array}
$$

Proof: This comes directly from naturality in Y of the bijections

$$\mathcal{P}(X \times Y) \simeq \mathcal{P}(Y \times X) \simeq \mathcal{E}(\mathcal{Y}, \mathcal{P}\mathcal{X})$$

     ∎

For each object X of $\mathcal{E}$ we also have an equality relation which is merely the subobject of $X \times X$ corresponding to the map $\Delta : X \to X \times X$ characterized by $\pi_1 \circ \Delta : \pi_2 \circ \Delta = 1_X$.

In a topos we also have a terminal object 1.

Definition 12. *The object* $\Omega = P(1)$ *is called the* object of truth values of $\mathcal{E}$.

We have

$$P(X) \simeq P(X \times 1) \simeq \mathcal{E}(X, \Omega).$$

The arrow $\phi : X \to \Omega$ corresponding to a subobject A of X is called the *characteristic function* of A. The subobject of $1, id_1 : 1 \to 1$, has a characteristic function denoted by

$$t : 1 \to \Omega$$

and called *true*.

In Set, Ω is $P(\{*\})$ and thus Ω is a set with two elements often written $\{0, 1\}$. In the category of presheaves Ω is given by

$$\Omega(X) = \{\text{subfunctors of } h_X\},$$

that is, the set of sieves on X. In the topos of sheaves on a topological space X, the sheaf Ω is given by

$$\Omega(U) = \{V | V \text{ open and } V \subseteq U\}.$$

If we think of truth values as being possible answer to a question, the question is no longer "yes or no ?" but 'Where ?".

As a corollary of Proposition 6 we have

Proposition 7. *Let A be a subobject of X. The characteristic function $\phi_A : X \to \Omega$ is the unique map such that the following square is a pullback.*

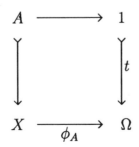

■

In a topos $\mathcal{E}$ we can form the intersection of two subobjects. It is defined by the pullback

$$
\begin{array}{ccc}
A \cap B & \longrightarrow & A \\
\downarrow & & \downarrow \\
B & \rightarrowtail & X
\end{array}
$$

We can also define an internal conjunction in Ω :

$$\wedge : \Omega \times \Omega \to \Omega$$

as the characteristic function of the subobject

$$(t,t) : 1 \to \Omega \times \Omega$$

It is trivial to see:

Proposition 8. *If* $A \to X$ *and* $B \to X$ *are two subobjects with characteristic map* ϕ_A *and* ϕ_B *then*

$$\phi_{A \cap B} = \phi_A \wedge \phi_B = \wedge \circ (\phi_A, \phi_B).$$

■

We now consider a subobject $\phi : A \rightarrowtail X_1 \times \cdots \times X_n$ and an index $i \, (1 \leq i \leq n)$. We write $Y = X_1 \times \cdots \times \hat{X}_i \times \cdots \times X_n$ for the product of these factors with index other than i. We can see A as a subobject of $Y \times X_i$ and consider the correponding arrow

$$Y \to P(X_i).$$

We denote this arrow by

$$|\{x \in X_i | \phi\}|.$$

This trivial but fundamental construction becomes more meaningful in the context of the language of the topos, to which we now turn.

8.2.4   Language and theory of a topos

$\mathcal{E}$ is a fixed topos. We associate to $\mathcal{E}$ a higher-order multisorted language $\mathcal{L}_\mathcal{E}$ in the sense of 8.1.1.

a) The class "Typ" will be the class $\mathcal{E}_0$ of objects of $\mathcal{E}$. We have the required constructions:

1. as constant 1 we take the terminal object.

2. as pairing map we take $E_0 \times E_0 \to E_0 : (X, Y) \to X \times Y$

3. as unary law we take $E_0 \to E_0 : X \to P(X)$

b) For each object X in E we take an infinite set Var_X of variables of type X. In order to avoid tedious manipulations with projections we will, by an abuse of language, identify variables of type $X \times Y$ with pairs (x, y) of variables of types X and Y.

c) For types X and Y we take

$$Op_{X,Y} = \mathcal{E}(X, Y)$$

and we have projections $\pi_1 : X \times Y \to X$ and $\pi_2 : X \times Y \to Y$.

d) For type X we take

$$Rel_X = P(X)$$

and we have seen in 8.2.3 that we have, for all X, the equality relation $=_X$ which is the diagonal morphism $X \to X \times X$ and the membership relation $\in_X \rightarrowtail X \times PX$.

e) Finally we have the logical symbol $\wedge$ and the collection symbol $\{\cdot|\cdot\}$.

Now we will define an interpretation of expressions (terms and formulas) of L_E.

First we fix a notation. If $\sigma = (x_1, \cdots, x_k)$ is a finite sequence of distinct variables of types $X_1, X_2, \cdots, X_k$ we put $X_\sigma = X_1 \times \cdots \times X_k$ and, in particular $X_\emptyset = 1$. If $\sigma' \subseteq \sigma$ we have an obvious projection from X_σ into $X_{\sigma'}$. Generally we denote such a projection by π.

Then, by induction, we define the interpretation $|E|$ of an expression E in such a manner that

a) If t is a term of type X, $|t|$ is an arrow $X_{\sigma(t)} \to X$

b) If ϕ is a formula, $|\phi|$ is a subobject of $X_{\sigma(\phi)}$.

It may be useful to consider more variables than those of the considered expression. If $\sigma(t) \subseteq \sigma$ we denote by $|t|^\sigma$ the composition of $\pi : X_\sigma \to X_{\sigma(t)}$ with $|t| : X_{\sigma(t)} \to X$ and by $|\phi|^\sigma$ the pullback of $|\phi|$ by $\pi : X_\sigma \to X_{\sigma(\phi)}$.

The induction is the following:

a) If x is of type X, $|x| = id_X$. If $\emptyset$ is the term of type 1, $|\emptyset| = 1$. If t_1 and t_2 are of types X_1 and X_2 we take

$$|(t_1, t_2)| = (|t_1|^{\sigma(t_1, t_2)}, |t_2|^{\sigma(t_1, t_2)})$$

which is an arrow from $X_{\sigma(t_1, t_2)}$ to $X_1 \times X_2$.

b) If t is a term of type X and $f : X \to Y$ is an arrow (i.e. an operation in $Op_{X,Y}$) we take $|f(t)| = f \circ |t|$

c) If $R \rightarrowtail X$ is a subobject of X (i.e. an element of Rel_X)and t is a term of type X, we define $|R(t)|$ by the pullback of R by $|t|$:

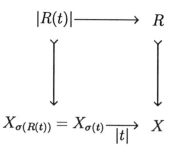

d) If ϕ and ψ are two formulas and $\sigma = \sigma(\phi \wedge \psi)$, we define $|\phi \wedge \psi| = |\phi|^\sigma \cap |\psi|^\sigma$.

e) Finally, let ϕ be a formula, and x a variable of type X. Given the interpretation of $|\phi| \rightarrowtail X_{\sigma(\phi)}$, we must describe the interpretation of $\{x|\phi\}$. We take $\sigma' = (\sigma(\phi) \backslash (x)) \cup (x)$ (a way to say that we add x to $\sigma(\phi)$ if x were not in $\sigma(\phi)$). We consider

$$|\phi|^{\sigma'} \rightarrowtail X_{\sigma'} = X_{\sigma(\phi) \backslash (x)} \times X.$$

By representation of binary relations in a topos, there is a unique corresponding morphism

$$X_{\sigma(\phi) \backslash (x)} \to PX$$

which we take to be $|\{x|\phi\}|$.

Now a formula ϕ is said to be valid in E, or E is said to be a model of ϕ, if $|\phi|$ is the identity on $X_{\sigma(\phi)}$.

Here a warning is necessary. A projection $X_\sigma \to X_{\sigma(\phi)}$ for $\sigma(\phi) \subseteq \sigma$ is not always an epimorphism. Some new factor may appear in X_σ which behaves like the emply set in <u>Set</u>. If $\pi : X_\sigma \to X_{\sigma(\phi)}$ is an epimorphism and $|\phi|$ is $1_{X_{\sigma(\phi)}}$ then $|\sigma|^\sigma$ is also 1_{X_σ}. But if π is not an epimorphism this conclusion does not follow. We write $\mathcal{E} \vDash \phi$ for $|\phi| = 1_{X_{\sigma(\phi)}}$ and we write $\mathcal{E} \overset{\sigma}{\vDash} \phi$ for $|\phi|^\sigma = 1_{X_{\sigma(\phi)}}$ (where $\sigma \supseteq \sigma(\phi)$).

We state the following results without proof.

Proposition 9. *Every intuitionistic theorem (as those of 8.1.4 - 8.1.5 - 8.1.6) is valid in any topos $\mathcal{E}$.* ∎

Proposition 10. *Every intuitionistic deduction rule is correct in any topos $\mathcal{E}$ with one restriction for the modus ponens, expressed by : if $\pi : X_{\sigma(\phi \Rightarrow \psi)} \to X_{\sigma(\phi)}$ is an epimorphism and $E \vDash \phi$ and $E \vDash (\phi \Rightarrow \psi)$ then $E \vDash \psi$.* ∎

As an example we point out that the correspondence between relations $R \to X \times Y$ and morphisms $\phi_R : Y \to PX$ given by Proposition 6 can be written as

$$R(x,y) \leftrightarrow x \in_X \phi_R(y).$$

This is a formula of set theory which is in fact valid in any topos. This is the precise meaning of the affirmation used in this book that we are working in a kind of intuitionistic set theory.

As another example of internal set-theoretic notion we point out the internal concept (i.e. object) of the "set" of morphisms from X to Y. Let X and Y be two objects in a topos $\mathcal{E}$ and r be a variable of type $P(X \times Y)$. We can consider the formula

$$\forall x \, \exists y \, (x,y) \in r \wedge \forall x \, \forall y \, \forall y' \, ((x,y) \in r \wedge (x,y') \in r \to y = y').$$

In this formula the only free variable is r . The interpretation of this formula is a subobject of $P(X \times Y)$. It is internally the "set" of functional, and everywhere defined relations. It is called the object of morphisms from X to Y, and be denoted by Y^X. With the logic just described it is easy to show, just as in sets, that

$$\mathcal{E}(X \times Y, Z) \simeq \mathcal{E}(X, Z^Y).$$

In a topos of varying sets, Set^C, let F and G be two objects. We can define $G^F : C \to Set$ by $G^F(A) = Nat(h^A \times F, G)$ for $A \in C_0$ and $G^F(f) : G^F(A) \to G^F(B)$ is given by composing $\phi : h^A \times F \to G$ with $h^f \times F : h^B \times F \to h^A \times F$. The logical and categorical definitions of G^F agree in that topos.

In the case of a topos of varying sets or a topos of presheaves Set^{C^o}, it is possible to give another definition of the validity of a formula. We treat the case of presheaves (but the case of varying sets is dual).

Let ϕ be a formula, $\sigma(\phi)$ the finite sequence of its distinct variables and $\sigma \supseteq \sigma(\phi)$. Let F_σ be the corresponding object in the topos Set^{C^o}, A an object of C. An element of F_σ at stage A is an arrow $a : h_A \to F_\sigma$. As $h : C \to Set^{C^o} : A \mapsto h_A$ is an embedding of C in Set^{C^o} we may write A for h_A. We shall also write $a \in_A F_\sigma$ for $a : h_A \to F_\sigma$. If x is a variable of type F we often write $x \in F$.

Now we inductively define a relation denoted by

$$A \Vdash \phi[a]$$

and called *forcing*. This is a kind of satisfaction.

a) If R is in $Rel_n(F)$, and $a \in_A F$, we put $A \Vdash R[a]$ if and only if the interpretation of $R(a)$ is the identity on A.

b) The definition then proceeds inductively:

1. $A \Vdash (\phi \wedge \psi)[a]$ iff $A \Vdash \phi[a]$ and $A \Vdash \psi[a]$.

2. $A \Vdash (\phi \vee \psi)[a]$ iff $A \Vdash \phi[a]$ or $A \Vdash \psi[a]$

3. $A \Vdash \neg\phi[a]$ iff for all B and all $u : B \to A, B \nVdash \phi[a \circ u]$

4. $A \Vdash \forall x \in F\phi(x, a)$ iff for all B, all $u : B \to A$, and all $b \in_B F$, we have $B \Vdash \phi(b, a \circ u)$.

5. $A \Vdash \exists x \in F\phi(x, a)$ iff there exists $b : A \to F$ such that $A \Vdash \phi(b, a)$.

For example, it is easy to verify that

(1) For terms t and s of signature (F, G) and for $a \in_A F$, we have $A \Vdash t = s[a]$ if $t(a) =_G s(a)$.

(2) If t is a term of signature (F, G), τ is a term of signature (F, PG), and $a \in_A F$, then $A \Vdash (t \in \tau)[a]$ iff $t(a) \in_A \tau(a)$.

These proofs and the proof of the following proposition are relatively easy and can be found for example in Borceux [5] (Vol 3, section 6.6).

Proposition 11. *In a topos of presheaves* Set^{C^o}*, a formula* $\phi(x)$ *where* $x \in F$*, is valid if and only if, for all A in C and $a \in_A Ff$, $A \Vdash \phi[a]$.* ∎

There is a similar notion of forcing in toposes of sheaves. For simplicity and shortness we only describe the case of sheaves on a topological space. The slogan is "existence is local existence".

Let F be a sheaf on a topological space X and U an open subset of X. As for presheaves we write $a \in_U F$ for $a : h_U \to F$. The definition is given by

a) If R is in $Rel_n(F)$, and $a \in_U F$, we put $U \Vdash R[a]$ if and only if the interpretation of $R(a)$ is the identity on $F(U)$.

b) The definition proceeds inductively:

1. $U \Vdash (\phi \wedge \psi)[a]$ iff $U \Vdash \phi[a]$ and $U \Vdash \psi[a]$.

2. $A \Vdash (\phi \vee \psi)[a]$ iff there exists an open cover $(U_i)_{i \in I}$ of U and for each i in I, we have $U_i \Vdash \phi[a \, |_{U_i}]$ or $U_i \Vdash \psi[a \, |_{U_i}]$

3. $U \Vdash \neg\phi[a]$ iff for all $V \subseteq U$, we have $V \nVdash \phi[a \, |_V]$

4. $U \Vdash \forall x \in F \phi(x, a)$ iff for all $V \subseteq$ and all $b \in_V F$, $V \Vdash \phi(b, a \, |_V)$.

5. $U \Vdash \exists x \in F \phi(x, a)$ iff there exists an open cover $(U_i)_{i \in I}$ of U and for each i in I there exists $a_i \in_{U_i} F$ such that $U_i \Vdash \phi(a_i, a \, |_{U_i})$.

In the general case of a topos of sheaves we simply replace open covers by sieves in the localizing system.

8.3   Some models of S.D.G.

8.3.1   The algebraic model

The first example of a topos in which we have a model of the line R with Kock-Lawvere axiom, is an elementary one in fact too simple to be really useful in differential geometry. It is the topos of varying sets on the category $\mathbb{R} - alg_f$ of associative, commutative, unitary $\mathbb{R}$-algebras of finite type (in the usual sense i.e. with a finite number of genera- tors). They are quotient of a $\mathbb{R}$-algebra of polynomials $\mathbb{R}[X_1, \ldots, X_n]$ by an ideal. This situation is too simple but it permits to see how it is possible to think a kind of set theory in which all functions from R to R are differentiable. As a matter of fact, all these functions are poly- nomial and that is precisely why this topos is too simple for differential geometry. In that topos $Set^{\mathbb{R} - alg_f}$, objects are functors and morphisms are natural transformations.

Definition 1. *The object of line type ,written R, is the forgetful func- tor associating to each object A of $\mathbb{R} - alg_f$ the set of its elements and to each homomorphism $f : A \to B$, f itself as a function.*

Proposition 1. *In $Set^{\mathbb{R} - alg_f}$, R is a commutative and unitary ring.*

Proof: For each object A of $\mathbb{R} - alg_f$ (or each stage A as we can equivelently say), $R(A)$ is a ring. Thus the additive and muliplicative operations exist :

$$+_A : R(A) \times R(A) \to R(A)$$

$$\cdot_A : R(A) \times R(A) \to R(A)$$

These are natural because the arrows in $\mathbb{R} - alg_f$ are homomorphisms and so we have operators $+ : R \times R \to R$ and $. : R \times R \to R$. To verify the axioms, it suffices to look at each stage A where they are obviously true as $R(A)$ has a ring structure.   ∎

Proposition 2. *R is a representable functor (cf. 8.2.1 Definition 4).*

Proof: For each $\mathbb{R}$-algebra A we have, in a natural manner,

$$R(A) \simeq \mathbb{R} - alg(\mathbb{R}[X], A).$$

In fact to $a \in R(A)$ we associate the $\mathbb{R}$-homomorphism $\phi_a : \mathbb{R}[X] \to A$ characterized by $\phi_a(X) = a$. We can write $R = h^{\mathbb{R}[X]}$.   ∎

From an external point of view, maps from R to R are just polynomials :

$$
\begin{aligned}
\mathbb{R} - alg(R, R) &= Nat(h^{\mathbb{R}[X]}, h^{\mathbb{R}[X]} \\
&\simeq \mathbb{R} - alg(\mathbb{R}[X], \mathbb{R}[X]) \\
&\simeq \mathbb{R}[X].
\end{aligned}
$$

But we consider here what happens from the internal point of view.

As we said at the end of Section 8.2, in a topos $E = Set^C$ of varying sets, the internal object of morphisms G^F is described at stage A by

$$G^F(A) \simeq Nat(h^A \times F, G).$$

Proposition 3. *R^R is isomorphic to the functor associating to an $\mathbb{R}$-algebra A the set of elements of $A[X]$ (the $\mathbb{R}$-algebra of polynomials with coefficients in A).*

Proof: At stage A we have

$$
\begin{aligned}
R^R(A) &= Nat(h^A \times R, R) \\
&= Nat(h^A \times h^{\mathbb{R}[X]}, h^{\mathbb{R}[X]}) \\
&= Nat(h^{A \otimes \mathbb{R}[X]}, h^{\mathbb{R}[X]}) \\
&= \mathbb{R} - alg(\mathbb{R}[X], A \otimes \mathbb{R}[X]) \\
&= A \otimes \mathbb{R}[X] \\
&= A[X]
\end{aligned}
$$

and naturality in A is straightforward   ∎

Functors represented by Weil-$\mathbb{R}$-algebras correspond to small objects in $Set^{\mathbb{R}-alg}$. For example:

$$D : \mathbb{R} - alg \to Set : A \mapsto D(A) = \{a \in A | a^2 = 0\}$$

is in fact $h^{\mathbb{R}[\epsilon]}$ and more generally this is the case for every Weil-algebra (cf.2.1). Thus we have the following proposition.

Proposition 4. *Objects R and D satisfy the Kock-Lawvere axiom. More generally for any Weil-algebra W we have the (K-W) axiom (cf. 2.2.2).*

Proof: We restrict ourselves to the particular case of D. At stage A :

$$
\begin{aligned}
R^D(A) &= Nat(h^A \times D, R) \\
&= Nat(h^A \times h^{\mathbb{R}[\epsilon]}, h^{\mathbb{R}[X]} \\
&= \mathbb{R} - alg(\mathbb{R}[X], A \otimes \mathbb{R}[\epsilon]) \\
&= A \otimes \mathbb{R}[\epsilon].
\end{aligned}
$$

∎

Thus we have an example of a topos with the general (K-W)-axiom. It is even possible to have a weak form of the integration axiom. Briefly, the concepts of derivation and of integration work as in polynomial algebras since functions from R to R are exactly polynomials. But, of course, this is not a good model. First, the integration axiom on $[0, 1]$ does not make sense since the interval $[0, 1]$ is not available. But more fundamental is the fact that the only manifolds that we can hope to find in this topos are algebraic ones and they are not enough for differential geometry.

8.3.2   Varying sets on $\mathcal{C}^\infty$-algebras

In the formal description of the theory of $\mathbb{R}$-algebras, the operations of addition, of multiplication, and of product by a scalar are usually privileged and one has axioms expressing fundamental equalities between certain composites of these operations (as for example $\alpha(x+y) = \alpha x + \alpha y$). But it can be instrucive to be less economical and to consider as n-ary operations all the polynomials in $\mathbb{R}[X_1, \ldots, X_n]$ and as axioms all the equalities between composites of polynomials which are true in the reals. A $\mathbb{R}$-algebra thus is a model of this theory of $\mathbb{R}$-polynomials.

We shall extend this situation, substituting $\mathcal{C}^\infty$-functions for polynomials.

Definition 2. *The theory of C^∞- algebras is the theory having as n-ary operations all the functions in $C^\infty(\mathbb{R}^n, \mathbb{R})$ and as axioms all equalities between composites of C^∞-functions which are true in the reals.*

Definition 3. *A C^∞-algebra is a model of the theory of C^∞-algebras. In other words, for each function $f \in C^\infty(\mathbb{R}^n, \mathbb{R})$, we have an interpretation $A(f) : A^n \to A$ in such a manner that, if $h = g \circ (f_1, \dots, f_k)$ (where $f_i \in C^\infty(\mathbb{R}^n, \mathbb{R}), g \in C^\infty(\mathbb{R}^k, \mathbb{R})$ and $h \in C^\infty(\mathbb{R}^n, \mathbb{R})$) then $A(h) = A(g) \circ (A(f_1), \dots, A(f_k))$.*

Now here are some examples of C^∞-algebras.

1) $C^\infty(\mathbb{R}^k, \mathbb{R})$ is a C^∞-algebra. If $f \in C^\infty(\mathbb{R}^n, \mathbb{R})$, it is interpreted by composition:

$$C^\infty(f) : C^\infty(\mathbb{R}^k, \mathbb{R})^n \to C^\infty(\mathbb{R}^k, \mathbb{R})$$

$$(g_1, \dots, g_n) \mapsto f \circ (g_1, \dots, g_n)$$

As $\mathbb{R}[X_1, \dots, X_k]$ is the free $\mathbb{R}$-algebra with k generators, in the same way $C^\infty(\mathbb{R}^k, \mathbb{R})$ is the free C^∞-algebra with k generators.

2) We recall that a function $f : X \to \mathbb{R}$ ($X \subseteq \mathbb{R}^n$) is said to be a C^∞-function if it is the restriction to X of a C^∞-function on $\mathbb{R}^n$ (or an open subset of it containing X. The set $C^\infty(X, \mathbb{R})$ of C^∞-functions from X to $\mathbb{R}$ is a C^∞-algebra.

3) If M is a manifold of class C^∞ (in the usual sense), $C^\infty(M, \mathbb{R})$ is a C^∞-algebra. If $f \in C^\infty(\mathbb{R}^n, \mathbb{R})$ we, of course, define

$$C^\infty(f) : C^\infty(M, \mathbb{R})^n \to C^\infty(M, \mathbb{R})$$

by composition as in example (1).

4) If A is a C^∞-algebra and I is an ideal in A, A/I also is a C^∞-algebra. It suffices to see that if $f \in C^\infty(\mathbb{R}^n; \mathbb{R}), A(f) : A^n \to A$ passes to the quotient, i.e. if $\underline{a}$ and $\underline{b}$ are equivalent modulo I in A^n then $A(f)(\underline{a})$ is equivalent modulo I to $A(f)(\underline{b})$. But, by Taylor's formula, there exist functions $g_i : \mathbb{R}^n \times \mathbb{R}^n \to \mathbb{R}$ such that

$$f(\underline{x}) - f(\underline{y}) : \sum_{i=1}^{n}(x_i - y_i)g_i(\underline{x}, \underline{y})$$

and a well known result of Hadamard shows that if f is C^∞ then so are the g_i .Thus A must preserve the above identity and we have:

$$A(f)(\underline{a}) - A(f)(\underline{b}) = \sum_{i=1}^{n}(a_i - b_i)A(g_i)(\underline{a}, \underline{b})$$

and thus, as a_i is equivalent to b_i modulo I, we have the result.

5) We know (cf Section 1.2) that a real Weil-algebra can be presented in the form $\mathbb{R}[X_1,\ldots,X_n]/I$ where I is an ideal generated by a finite number of polynomials such that for m big enough the X_i^m are in I. We now consider a C^∞-function f. We can define an interpretation of f in $\mathbb{R}[X_1,\ldots,X_n]/I$ as the interpretation of the Taylor polynomial of order k big enough (for example $k = n.m$). So we see that each real Weil-algebra is a C^∞-algebra.

6) It is easy to verify that the $\mathbb{R}$-algebra $\mathbb{R}[[X_1,\ldots,X_n]]$ of formal power series is also a C^∞-algebra.

It is interesting to consider the category of C^∞-algebras which are of finite type, i.e. those algebras of the form $C^\infty(\mathbb{R}^n)/I$. It is easy to see, from classical analysis, that, for X closed in $\mathbb{R}^k$, $C^\infty(X, \mathbb{R})$ is of finite type : it is $C^\infty(\mathbb{R}^n)/I$ where I is the ideal of C^∞-functions with value 0 on X. Here are two consequences of this fact.

a) If U is open in $\mathbb{R}^n$, from classical analysis we know that there exists a function f in $C^\infty(\mathbb{R}^n)$ such that $f(x) = 0$ iff $x \notin U$. Then U is diffeomorphic to the closed set $X = \{(x,y)|y \cdot f(x) = 1\}$ in $\mathbb{R}^{n+1}$ and so $C^\infty(U, \mathbb{R})$ is of finite type.

b) Every paracompact C^∞-manifold M can be embedded as a closed set in some $\mathbb{R}^n$. Thus $C^\infty(M, \mathbb{R})$ is also of finite type.

In fact we have more. A C^∞-algebra of finite type $C^\infty(\mathbb{R}^n)/I$ is said to be finitely presented if the ideal I is of finite type. For example if U is open in $\mathbb{R}^n$ we have just seen that $C^\infty(U, \mathbb{R}) \simeq C^\infty(\mathbb{R}^{n+1})/I$ where I is the ideal generated by one function $y \cdot f(x) - 1$ where f is in $C^\infty(\mathbb{R}^n)$ and $f(x) = 0$ iff $x \notin U$. So $C^\infty(U, \mathbb{R})$ is finitely presented.

But each paracompact manifold is a retract of an open subset in some $\mathbb{R}^n$. So $C^\infty(M, \mathbb{R})$ is also finitely presented. We have a good (contravariant) representation of C^∞-manifolds in C^∞-algebras.

Proposition 5. *The contravariant functor of the category Man^∞ of paracompact C^∞-manifolds to the category C^∞-Alg$_{fp}$ or C^∞-Alg$_{fg}$ of* finitely presented (or finitely generated) C^∞-algebras, sending M to $C^\infty(M, \mathbb{R})$ and $\phi : M_1 \to M_2$ to $C^\infty(\phi) : C^\infty(M_2) \to C^\infty(M_1)$ (defined by composition), is full and faithful.

For a proof (easy) we refer the reader to the book of I. Moerdijk and G.E. Reyes (op.cit.). There, it is also proved (less easily) that transversal pullbacks are sent to pushouts.

Now, to obtain a topos, we consider the topos of varying sets on $C^\infty\text{-}Alg_{fg}$:

$$S = Set^{C^\infty - Alg_{fg}}$$

(we write S for "smooth"). If we consider the (contravariant) Yoneda embedding:

$$C^\infty - Alg_{fg} \to S$$

$$A \mapsto C^\infty(A, -)$$

we obtain a covariant embedding of Man^∞ into S which is full and faithful and preserves transversal pulbacks.

With the same proofs as in the algebraic case, we have the following results.

Proposition 6.

a) in S the forgetful functor $R : C^\infty\text{-}Alg_{fg} \to Set : A \mapsto \{x | x \in A\}$ is represented by $(\mathbb{R})$ i.e. $R(A) \simeq C^\infty\text{-}Alg(C^\infty(\mathbb{R}), A)$. It is a commutative unitary ring.

b) The functor

$$D = C^\infty - Alg(C^\infty(\mathbb{R})/(x^2), -)$$

is, internally in S , just $\{d \in \mathbb{R} | d^2 = 0\}$. And we have the Kock-Lawvere axiom.

c) For all Weil-algebras we also have the (K-W)-axiom.

■

We also have, as in the algebraic case, a description of function spaces ("exponentials"). If F and G are in $\mathcal{S}$, G^F is described at stage A by $G^F(A) = Nat(h^A \times F, G)$.

But in $\mathcal{S}$ we have more objects than in the algebraic topos. For example, the interval $[0, 1]$ is the representable functor

$$\mathcal{C}^\infty\text{-}Alg(\mathcal{C}^\infty(\mathbb{R})/I, -)$$

where I is the ideal of $\mathcal{C}^\infty$-functions with value 0 on the interval $[0, 1]$ in $\mathbb{R}$. In the same way we can describe the "set" of positive reals. Now we have

Proposition 7. *The order axioms and the integration axiom are valid in the topos $\mathcal{S}$.*

Proof: Cf. Moerdijk and Reyes (op. cit.).   ∎

8.3.3   A good model of S.D.G.

The topos $\mathcal{S} = Set^{\mathcal{C}^\infty - Alg_{fg}}$ is in some sense a model too big and too rough. Here are some failings of this model.

a) The category Man^∞ of paracompact $\mathcal{C}^\infty$-manifolds has some interesting colimits, for example union of open coverings, which are not preserved by the embedding of Man^∞ into $\mathcal{S}$.

b) The smooth line R is not a local ring (a local ring is a ring in which for all x, either x or $1 - x$ is invertible).

c) The smooth line R is not Archimedean. (This defect also can be seen as an asset because it leave to hope for a topos in which we have a model of synthetic differential geometry but also of non standard analysis. This is studied in Moerdijk and Reyes, op. cit., chapter VI).

We have two possibilities at our disposal

(i) We may limit ourself to $\mathcal{C}^\infty$-algebras $\mathcal{C}^\infty(\mathbb{R}^n/I)$ for some particular kind of ideals.

(ii) Instead of presheaves we can limit ourself to sheaves for a Grothendieck localising system in the category of $\mathcal{C}^\infty$-algebras used.

We point out only one example where the indicated defects disappear. This example is thought to be the most useful in differential geometry.

Some preliminary definitions are convenient.

Definition 4. *Let x be in $\mathbb{R}^n$. We say that two functions u and v defined on open neighborhoods of x have the same germ at x if they coincide on a neighborhood of x.*

We denote by $f|_x$, and we call the germ of f at x, the equivalence class of functions having the same germ as f at x. The set of germs at x is denoted by $C_x^\infty(\mathbb{R}^n)$ and if I is an ideal $I|_x$ denotes the set of germs at x of elements of I. It is an ideal of $C_x^\infty(\mathbb{R}^n)$.

Definition 5. *Let I be an ideal of $C^\infty(\mathbb{R}^n)$. We define the set of zeros of I, written $Z(I)$, by :*

$$Z(I) = \{x \in \mathbb{R}^n | \forall f \in I \ f(x) = 0\}.$$

Definition 6. *We say that I is germ-determined if for all f in $C^\infty(\mathbb{R}^n)$ we have that if for each x in $Z(I)$, $f|_x \in I|_x$ then f is in I.*

We consider the category $G = C^\infty\text{-}Alg_{gd}$ of C^∞-algebras of type $C^\infty(\mathbb{R}^n)/I$ where I is a germ-determined ideal.

It can easily be seen that an ideal I of $C^\infty(\mathbb{R}^n)$ is germ-determined if and only if for all locally finite families $(f_\alpha)_\alpha$ of elements of I, Σf_α also is in I. In particular a finitely generated ideal is germ determined.

If I is an ideal in $C^\infty(\mathbb{R}^n)$, there exists a smallest germ-determined ideal $\tilde{I}$ containing I. We take $\tilde{I}$ to be the set of the Σf_α for all locally finite families (f_α) of elements of I. Also $\tilde{I}$ is the ideal comprising exactly those C^∞-functions f such that for all x in $Z(I)$, $f|_x \in I|_x$.

It must be observed that if I is germ-determined in $C^\infty(\mathbb{R}^n)$ and U is an open in $\mathbb{R}^n$, $I|_U$ is not always germ-determined ; one has to consider $\widetilde{I|_U}$.

We will now describe a Grothendieck localizing system on G^o (the dual of G) and consider the topos of sheaves on G^o for that system.

Let $A = C^\infty(\mathbb{R}^n)/I$ be an object of G. A covering family for A is a family of morphisms of C^∞-algebras:

$$(u_\alpha : C^\infty(\mathbb{R}^n)/I \to C^\infty(U_\alpha)/\widetilde{I|_{U_\alpha}})_\alpha$$

where $(U_\alpha)_\alpha$ is an open cover of $Z(I)$. Such a family determines a subfunctor of $G(A, -)$ which to an object B of G associates the set of morphisms from A to B in G having a factorization through one of the u_α. Of course, in the dual G^o, we obtain a subfunctor of the contravariant functor $G^o(-, A)$ on G^o.

In fact it is easy to prove that we thus obtain a Grothendieck localizing system on G^o. The desired topos is the topos $\mathcal{G}$ of sheaves on G^o for that system.

In Moerdijk and Reyes (op. cit.) the following results can be found.

a) Representable presheaves $G^o(-, A)(= G(A, -))$ are sheaves for the indicated Grothendieck localizing system.

b) The category Man^∞ of paracompact C^∞-manifolds has a full and faithful covariant embedding in $\mathcal{G}$ which preserves transversal pullback and also transforms open coverings into covering families.

c) In $\mathcal{G}$ we have the Kock-Lawvere axiom and the general Kock-Weil axiom for R (the forgetful functor which is representable and thus is a sheaf) . We also have the order axioms and the integration axiom.

d) In $\mathcal{G}$ the smooth line R is a local ring.

e) In $\mathcal{G}$, R is Archimedian.

So the failings of the topos $\mathcal{S}$ mentioned above, disappear in $\mathcal{G}$, where we have a good model of synthetic differential geometry.

8.4   Commented bibliography

In these lectures we have not tried to treat comprehensively the question of models of S.D.G. This chapter is only introduction. The introduction to categories is written for the reader without any knowledge of the subject. The introduction to toposes is more important but the interested reader will, of course, have to read more advanced books on that subject. We quote, for example, that of P.T. Johnstone [23]. Here our aim is only to introduce one model of S.D.G.

If we are interested only in logical coherence of the theory, from the intuitionistic point of view, we can consider the simple topos Set^A, which is like a universe in which we can give a meaning to the intuitionistic logic of superior order. But, in addition, the forgetful functor from A to Set, denoted R, satisfies the K-W axioms. So we have a rather elementary model of synthetic differential geometry.

This type of model is already suggested by Lawvere [44] and Kock [27] (1977). It is, however, only logically and not geometrically satisfactory in the sense, as we have seen, that it is too strict : in it all functions from ℝ to ℝ appear to be polynomial and not just "C^∞".

It was Dubuc who constructed a model E in which the category of C^∞-differentiable paracompact finite dimensional manifolds is immersed in an entirely accurate way. These questions have been dealt with in books of Kock [31] (1981), and Moerdijk and Reyes [52] (1991). The original articles of Dubuc on this matter are [14] (1979) and [12] (1981). On the question of models the book of Moerdijk and Reyes [52] is the most important reference.

BIBLIOGRAPHY

[1] ABRAHAM, R. , MARSDEN, J. E. *Foudations of Mechanics*, Benjamin, 1978.

[2] BELAIR,,L., *Calcul infinitésimal en géométrie différentielle synthétique*, Master Thesis, Montréal, 1981.

[3] BELAIR, L., et REYES, G.E., *Calcul infinitésimal en géométrie différentielle synthétique*, Sydney Category Seminar Reports,1982, 32pp.

[4] BERGERON, F., *Objet infinitésimalement linéaire dans un modèle bien adapté de G.D.S.*, in Géométrie Différentielle synthétique, Section 2, Analysis in smooth topos, edited by G.E. REYES, Research report, Montréal, (1980), 67-76.

[5] BORCEUX, F., Handbook of Categorical Algebra , 1, Basic Category Theory, 2, Categories and Structures, 3, Categories of Sheaves. Cambridge University Press, 1994.

[6] BUNGE, M., *Synthetic aspects of C^∞-mappings*, J. Pure Appl. Algebra **28** (1983), 41-63.

[7] BUNGE, M., and DUBUC, E.J., *Archimedean local C^∞-rings and models of synthetic differential geometry*, Cahiers Top. Géom. Diff. Cat. **27** (1986), 3-22.

[8] BUNGE, M., and DUBUC, E.J., *Local concepts in synthetic differential geometry and germ representability*, in "Mathematical Logic and Theoretical Computer Sciences", D. Kueher, E.G.K. Lopez-Escobar and C. Smith eds., Dekker, New York, 1987, pp. 39-158.

[9] BUNGE, M. and GAGO, F., *Synthetic aspects of C^∞-mappings II: Mather's theorem for infinitesimally represented germs*, J. Pure Appl. Algebra **55** (1988), 213-250.

[10] BUNGE, M., and SAWYER, P., *On connections, geodesics and*

sprays in synthetic differential geometry, Cahier Top. Géom. Diff., **25** (1984), 221-258 (preprinted in ([26])).

[11] CARTAN, H., Notions d'algèbres différentielle; applications aux groupes de Lie et aux variétés où opère un groupe de Lie, Colloque de Topologie, Bruxelles, 1951, 15-27.

[12] DUBUC, E.J., C^∞ *schemes*, Amer. journ. Math. **103** (1981), 683-690.

[13] DUBUC, E.J., *Open covers and infinitary operations in* C^∞-*rings*, Cahiers Top. Géom. Diff., **22** (1981), 287-300.

[14] DUBUC, E.J., *Sur les modèles de la géométrie différentielle synthétique*, Cahiers Top. et Géom. diff. **20** (1979), 231-279.

[15] DUBUC, E.J., and PENON, J., *Objects compacts dans les topos*, J. Austr. Math. Soc., Ser. A, **40** (1986), 203-217.

[16] DUBUC, E.J., and REYES, G.E., *Subtoposes of the ring classifier*, in ([25]), 101-122.

[17] EHRESMANN, Ch., *Oeuvres complètes et commentées* I.1 et I.2, Suppléments 1 et 2 au vol. XXIV des Cahiers de Topologie et Géométrie Différentielle, (1983).

[18] FELIX, Y., and LAVENDHOMME, R., *On De Rham's theorem in synthetic differential geometry*, J. Pure Appl. Algebra **65** (1990), 21-31.

[19] GAGO, F., *Morse germs in SDG*, in "Categorial Algebra and its Applications", F. Borceux ed., Springer Lecture Notes 1348 (1988), 125-129.

[20] GAGO, F., *Singularités dans la géometrie différentielle synthétique*, Bull. Soc. Math. Belgique, Ser. A, **41** (1989), 279-287.

[21] GODBILLON, C., *Géométrie différentielle et mécanique analytique*, Paris, Hermann, 1969.

[22] GROTHENDIECK, A., in collaboration with J. DIEUDONNE, *Eléments de géométrie algébrique*, Inst. Hautes Et. Sc., publ. math. n°s 4, 8, 11, 17, 20, 24, 28, 32 (de 1960 à 1967).

[23] JOHNSTONE, P. T., *Topos theory*, Academic Press, 1977.

[24] KOBAYASHI, S., and NOMIZU, K., *Foundations of differential geometry*, Interscience, vol. I 1963, vol. II, 1969.

[25] KOCK, A.,(ed), *Topos theoretic Methods in Geometry*, edited by A. Kock, Various Publications Series n30, Aarhus, 1979.

[26] KOCK, A.,(ed), *Category theoretic Methods in Geometry*, edited by A.Kock, Various Publications Series n35, Aarhus, 1983.

[27] KOCK, A., *A simple axiomatic for differentiation*, Math. Scand. **40** (1977), 183-193.

[28] KOCK, A., *Taylor Series calculus for ring objects of line type*, Journ. Pure Appl. Alg. **18** (1978), 271-293.

[29] KOCK, A., *Differential forms in synthetic differential geometry*, Aarhus Preprint Series n28(1978/79), 11pp.

[30] KOCK, A., *On the synthetic theory of vector fields*, in [25], 139-157.

[31] KOCK, A., *Synthetic Differential Geometry*, London Math. Soc. Lect. Note Series 51, Cambridge Univ. Press, 1981.

[32] KOCK, A., *Properties of well adapted models for synthetic differential geometry*, Journ. Pure Appl. Alg. **20** (1981), 55-70.

[33] KOCK, A., *Differential forms with values in groups*, Bull. Australian Math. Soc. **25** (1982), 357-386.

[34] KOCK, A., *A combinatorial theory of connections*, Contemporary Mathematics, AMS, **30** (1984), 132-144.

[35] KOCK, A., and LAVENDHOMME, R., *Strong infinitesimal linearity, with applications to strong difference and affine connections*, Cahiers Top. Geom. diff. **25** (1984), 311-324.

[36] KOCK, A., and REYES, G.E., *Connections in formal differential geometry*, in [25], 158-195.

[37] KOCK, A., and REYES, G.E., *Models for synthetic integration theory*, Math. Scand. **48** (1981), 145-152.

[38] KOCK, A., REYES, G.E., and VEIT, B., *Forms and integratin in synthetic differential geometry*, Aarhus Preprint Series, n31 (1979/80).

[39] KOSZUL, J.-L., *Homologie et cohomologie des algèbres de Lie* Bull. Soc. Math. France 78 (1950), 65-127.

[40] LAVENDHOMME, R., *Notes sur l'algèbre de Lie d'un groupe de Lie en géomtrie différentielle synthétique*, Sém. Math. pure n11, Louvain-la-neuve, (1981), 19pp.

[41] LAVENDHOMME, R., *Leçons deGéométrie Différentielle Synthétique Naïve*, Monographies de mathématiques 3, Ciaco, Louvain-la-neuve, (1987),204pp.

[42] LAVENDHOMME, R., *Objets de Lie*, Bull. Soc. Math. Belgique, Ser. B, **43** (1991), 83-112.

[43] LAVENDHOMME, R., *Algèbres de Lie et Groupes microlinéaires*, Cahiers Top. et Géom. Diff.Cat.,bf 35-1 (1994) 29-47.

[44] F.W. LAWVERE, *Categorical Dynamics*, in [25], 1-28.

[45] LIBERMANN, P., and MARLE, C.-M, *Symplectic Geometry and Analytical Mechanics*, Reidel, 1987.

[46] MacLARTY, C., *Local and some global results in synthetic differential geometry*, in ([26]), 226-256.

[47] MINGUEZ, M.C., *Calculo diferencial sintetico y su interpretacion en modelos de prehaces*, PhD. Thesis, 1985.

[48] MINGUEZ, M.C., *Wedge product of forms in synthetic differential geometry*, Cahiers Top. Geom. Diff. Cat. **29** (1988), 59-66.

[49] MINGUEZ, M.C., *Some combinatorial calculus on Lie derivative*, Cahiers Top. Geom. Diff. Cat. **29** (1988), 241-247.

[50] I. MOERDIJK and G.E. REYES, *Cohomology theories in synthetic differential geometry*, in ([26]), 1-67.

[51] I. MOERDIJK and G.E. REYES, *De Rham's theorem in a smooth topos*, Report 83-20, Amsterdam (1983), 22pp.

[52] MOERDIJK, I., and REYES, G.E., *Models for Smooth Infinitesimal Analysis*, Springer, 1991.

[53] PENON, J., *De l'infinitésimal au local*, State PhD. Thesis, Paris, 1985, 191pp.

[54] REYES, G.E., and WRAITH, G.E., *A note on tangent bundle in a category with a ring object*, Math. Scand. **42** (1978), 53-63.

[55] SPIVAK, M., *A Comprehensive Introduction to Differential Geometry*, Publisher Perish, Berkeley, vol. I-V, 1979.

[56] WEIL, A., *Théorie des points proches sur les variétés différentiables*, in Conference Geom. Diff., Strasbourg, 1953, took up again in "Oeuvres scientifiques, collected papers", vol. II, Springer 1979, 103-109.

Index

abstraction symbol 270
action of a Lie group 248
action of a Lie object 191
adjoint of a left action 250
adjoint representation 252
affine automorphism 246
affine vector field 246
C^∞-algebra 305
almost complex connection 229
almost complex structure 226
almost Hermitian object 230
h-antisymmetric 243
arc of marked surface 117
axiom
    Bunge - 59
    Fermat-Reyes - 26
    general Kock - 42
    integration - 19
    Kock-Lawvere - 2
    Koszul - 182
    Lie-module - 182
    preorder - 17
    Wraith - 58
based vector field 254
Bianchi identity 200
boundary operator 108
breadth of a Weil algebra 36
Bunge axiom 59
C^∞-algebra 305

characteristic function 295
category 279
classical
    - differential form 104
    - global differential form 184
coadjoint representation 253
coequalizer 283
complex object 227
connecting map 257
connecting mapping 163
connection 143
    almost complex - 229
    - form 163
    - on a vector bundle 255
    global - 145
    linear - 192
    pointwise - 143
    Riemannian - 212
    symmetric - 155
contravariant functor 281
coproduct 283
covariant derivation 147
covariant derivation on a vector bundle 257
covariant differential 169
curvature 176
curvature form 174
degenerate chain 136
de Rham cohomology 116

derivation	76
derivation axiom	182
derivative	6
differential form	102
classical - -	104
classical global - -	184
integral of a - -	110
singular - -	102
singular global - -	184
diffusion	264
diffusion on a vector bundle	264
directional derivative	11
differential	12
distribution of dimension 2	99
epimorphism	282
equalizer	283
Euclidean	5
Euclidan vector bundle	66
extensionality	278
exterior differential	115
exterior product	124
Fermat-Reyes axiom	26
finite limit	50
finitely complete	283
p-form on a Lie object	187
formula	271
functor	280
general Kock axiom	42
germ	309
germ-determined ideal	309
global	
- connection	145
- presymplectic structure	216
- Riemannian structure	212
good finite limit	52
h-antisymmetric	243
h-gradient	217
h-orthogonal	242
Hadamard's lemma	20
Hamilton equations	267
Hamiltonian	220
locally Hamiltonian	220
Hamilton operator	221
height of a Weil algebra	36
Hermitian object	230
Hermitian structure	230
h-gradient	217
holomorph	229
homogeneous	15
infinitesimal n-chain	108
k-infinitesimally linear object	58
initial object	283
integral of a differential form	110
integration axiom	19
iterated tangent bundle	87
interior product	126
isomorphism	282
Killing vector field	244
Kock-Lawvere Axiom	2
Kaelherian object	232
Koszul's axiom	182
left-invariant	240
linear connection	192
Lie	
action of a - group	248
action of a - object	191
- R-algebra	73
- bracket	73
- derivative	76
- derivative in the direction of X	236
- derivative of a form	128
- group	239
- module axiom	182

- object 182
- operator 77
Liouville's field 255
Liouville 1-form 259
Liouville 2-form 259
localizing system 293
locally Hamiltonian 220
membership relation 270
microcube 87
microlinear 57
micro-square 87
modus ponens 272
monomorphism 282
natural transformation 281
object of line type 302
operation symbol 270
h-orthogonal 242
parallel transport 164
pointwise
   - connection 143
   - Riemannian structure 211
Poisson
   - bracket 221
   - map 247
   - object 222
   - structure 223
   - tensor 224
   - map 247
   - vector field 247
preorder axiom 17
pre-symplectic
   global − structure 216
   − structure 216
   pointwise − structure 216
predicate calculus 276
presheaf 291
product 283

pullback 284
quasi colimit 52
reflexive immersed manifold 85
reflexive object 81
relations are representable 287
relation symbol 270
representable functor 282
Riemannian
   - connection 212
   - structure 211
   - structure on a Lie object 212
   pointwise - structure 211
   global - structure 216
set of zeros of an ideal 309
sheaf 291
   - w.r. a localizing system 293
singular
   - differential form 102
   - global differential form 184
small object 36
spectrum 33
spray 151
Stokes formula 121
strong difference 92
subobject 284
symmetric connection 155
symplectic structure 220
tensor product 123
term 271
terminal object 283
theory of $\mathcal{C}^{\infty}$-algebras 305
topos 287
torsion 171
   - form 170
   - of an almost complex object 226

truth values object   278
  - - - (in a topos)   295
type   270
vector bundle   66
Euclidan vector bundle   66
variable   270
varying set   287
vector field   69
vector tangent   61
Weil
     - algebra   35
     - characteristic homom...210
     breadth of a - algebra   36
     height of a Weil algebra 36
well powered   284
Whitney sum   159
Wraith axiom   58